# What's that
# TREE?

# What's that
# TREE?

## Tony Russell

**LONDON, NEW YORK, MUNICH, MELBOURNE, AND DELHI**

**DK LONDON**
**Senior Art Editor** Spencer Holbrook
**Senior Editor** Angeles Gavira
**Project Editor** David Summers
**US Senior Editor** Rebecca Warren
**US Editor** Jill Hamilton
**Pre-production Producer**
Nikoleta Parasaki
**Producer** Alicia Sykes
**Jacket Designer** Mark Cavanagh
**CTS** Sonia Charbonnier
**Managing Art Editor** Michelle Baxter
**Publisher** Sarah Larter
**Art Director** Philip Ormerod
**Associate Publishing Director**
Liz Wheeler
**Publishing Director** Jonathan Metcalf

**DK DELHI**
**Deputy Managing Art Editor**
Sudakshina Basu
**Design Consultant** Shefali Upadhyay
**Managing Editor** Rohan Sinha
**Senior Art Editor** Anuj Sharma
**Senior Editor** Anita Kakar
**Designer** Sanjay Chauhan
**Editor** Suefa Lee
**Senior DTP Designer** Harish Aggarwal
**DTP Manager/CTS** Balwant Singh
**Production Manager** Pankaj Sharma

First published in the United States in 2013
by DK Publishing
345 Hudson Street, New York,
New York 10014

4 6 8 10 9 7 5 3
015 – 187844 – March/2013

A catalog record for this book is available
from the Library of Congress
ISBN 978-1-46540-219-6

DK books are available at special discounts
when purchased in bulk for sales promotions,
premiums, fund-raising, or educational use.
For details contact: DK Publishing Special
Markets, 345 Hudson Street, New York, 10014
or SpecialSales@dk.com.

Printed and bound in China by
South China Co. Ltd.

Discover more at
**www.dk.com**

## ABOUT THE AUTHOR

A forester, broadcaster, and author, **Tony Russell** has worked in the field of trees and plants for 30 years, and is regarded as one of Europe's leading authorities on the subject. He joined the UK Forestry Commission in 1978, and in 1983 he was appointed Forester in the New Forest in Hampshire. In 1989, he was appointed Head Forester of the United Kingdom's National Arboretum at Westonbirt, Gloucestershire, which is considered one of the finest collections of trees and shrubs in the world. He has worked on numerous books on trees and regularly makes appearances on television and radio. He is author of *Smithsonian Nature Guide: Trees*.

# Contents

# Introduction

Trees are the largest, oldest, and most complex plants on Earth. They have been around for over 350 million years and cover almost one-third of the Earth's land surface. There are more than 80,000 different species (plus numerous cultivars), ranging in size from tiny Arctic Willows, just a few inches high, to Giant Redwoods over 300 feet (100 meters) tall.

This book is broken down into conifers, simple broadleaves, and compound broadleaves and then further subdivided by leaf shape. It features the trees that are most commonly found in towns, parks, and gardens, and in the surrounding countryside. Each species is described using clear, concise text and high-quality digital photography and illustrations. Where relevant, information is given on seasonal variations, such as the production of flowers or fruit, changing leaf color, and winter twigs and buds. This book will enable you to identify, understand, and enjoy more than 150 different trees, and will hopefully awaken a lifelong interest.

Tony Russell

# Identifying Trees

Looking at and identifying trees can be a fascinating and enjoyable pastime. However, unless you know what to look for, it can sometimes be confusing. The following pages aim to provide information to aid accurate identification. Two things to always bear in mind are season and location. These can affect how a tree grows and what it looks like.

## Season

Many trees display seasonal differences. Deciduous trees that are in full leaf in summer will have bare branches in winter. Deciduous leaf color in spring will be very different from the color in the fall. Features like winter twigs and buds, fall leaves, and spring flowers can be helpful in identifying trees in different seasons.

**FALL LEAVES**

**WINTER TWIG**

## Location

The forms of trees can vary by location. For example, trees growing close to each other will be taller and thinner, with fewer low branches than trees growing on their own in an open setting. Trees in exposed or cold locations are unlikely to grow as tall as those in sheltered locations.

**SHELTERED HILLSIDE**

**EXPOSED MOUNTAINSIDE**

# Tree Habit

Sometimes trees can be identified from a distance simply by their habit, or form. Many conifers will have a tall conical appearance, whereas many broadleaves will be broadly spreading. Just looking at the form of a tree should enable you to discount a number of identification possibilities; once discounted, further clues, such as leaf shape, can be used for identification.

## Form

Several terms are used to describe a tree's form. For example, "spreading" means the natural growth of the branches is directed away from the main trunk. "Columnar" indicates that the branches stay close to the trunk. "Shrublike" refers to a tree which has multiple branches or stems that arise from the base. "Conical" means the branches grow progressively shorter the higher they are on the tree.

SPREADING

COLUMNAR

SHRUBLIKE

CONICAL

## Size and growth rate

Trees display variability in their size and growth rate. For example, some oaks grow taller and faster than others of the same species due to factors such as location, soil, and the availability of water and sunlight. However, different tree species may have vastly different heights and growth rates. For example, an Arctic Willow may grow less than ½ in (1 cm) in 10 years, whereas the Lawson Cypress may grow more than 39 in (100 cm) in a year.

### Giant Redwood

The Giant Redwood, also called Wellingtonia, reaches heights in excess of 330 ft (100 m) in its native California, USA.

# Leaf Shapes

One of the best ways to identify a tree is by its leaves. Studied closely, they offer certain unique characteristics that can lead to a positive identification without referring to the tree's other characteristics (although it is always best to check these for confirmation). A leaf's veining, margin, stalk, color, sheen, and hair-covering are all important clues that aid identification.

### Long and thin
Also known as linear, this leaf type is usually associated with evergreen conifers.

### Needlelike
Found on most pines, firs, and spruces, these leaves may have a sharp or blunt point.

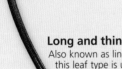

### Scalelike
Tiny flattened leaves that overlay one another.

### Heart-shaped
Also known as cordate, these leaves have a deep heart-shaped indent at the base.

### Egg-shaped
Also known as ovate, these leaves are broadest below the middle, like an egg.

### Inversely egg-shaped
Also known as obovate, these leaves taper at both ends but are broadest above the middle.

### Lance-shaped
Also known as lanceolate, these leaves widen just above the base before tapering toward the tip.

### Elliptic
Widest at about the middle, this leaf type narrows equally toward both ends.

### Rounded
This leaf shape, also known as orbicular, is almost circular in outline.

### Triangular
Also known as deltoid, these leaves are most commonly found on poplar trees.

### Oblong
These leaves are invariably longer than they are broad and have parallel, or nearly parallel, sides.

### Lobed
This distinctive leaf shape is common to most members of the oak family.

### Hand-shaped
Also known as palmate, these leaves are lobed in a handlike way. They are commonly found on maples.

### Fernlike
These long, feathery leaves are made up of many tiny leaflets.

### Divided
Also known as compound or pinnate, these leaves are divided into smaller leaflets.

### Twice-divided
These leaves have divided leaflets and are known as bipinnate.

### Trifoliate
These leaves consist of three leaflets, and are commonly seen on laburnums.

### Hand-shaped, divided
Also known as compound palmate, these leaves are seen on the Horse Chestnut.

# Leaf Margins

In addition to the shape, the edges or margin of the leaf can also aid in identification. Some trees, such as oaks, have leaves with lobes around the margin. Others, such as hollies, have leaves with spiny margins. The margins of some leaves may also be toothed to varying degrees, from double-toothed to serrated and saw-toothed.

SMOOTH

WAVY

TOOTHED

LOBED

DEEP-LOBED

SPINY

SAW-TOOTHED

SERRATED

DOUBLE-TOOTHED

# Leaf Arrangement

Once you have determined the shape and margin of a leaf, it is important to note how it is arranged on the shoot. Different trees can have virtually identical leaves, but their arrangement may be completely different. For example, Dawn Redwood leaves are borne opposite each other, whereas the very similar leaves of the Swamp Cypress are borne alternately along the shoots.

## Single leaves

Leaves can be borne singly on a shoot, rather than in groups or sprays. Conifers such as redwoods, firs, and spruces fall into this category, as do broadleaved trees such as hollies.

**DAWN REDWOOD**

## Leaves in clusters

Leaves can be borne in clusters or groups. The Scots Pine carries two needlelike leaves within a single basal sheath, which is attached to the shoot, while the Bhutan Pine carries five.

**BHUTAN PINE**

## Flattened sprays

Members of the cypress family have tiny, flattened leaves, which lie across each other to produce dense sprays of soft foliage that tend to droop slightly.

**LEYLAND CYPRESS**

## Opposite leaves

Many trees bear their leaves in pairs, each one opposite the other on a shoot. This regimented looking arrangement is apparent in rowans and maples.

**ROWAN**

## Alternate leaves

Some leaves are borne alternately along a shoot. Such a leaf arrangement gives a zig-zag appearance to the shoot, as in the case of the Caucasian Elm.

**CAUCASIAN ELM**

# Bark

Bark is the corky, waterproof layer that protects a tree's living tissue against disease and external attack. It has millions of tiny breathing pores, called lenticels, that pass oxygen from the atmosphere into the tree. Constantly visible and easily accessible, bark is a useful tool for identification. Trees can be distinguished by their bark color, texture, and markings.

### Peeling in strips

The Himalayan Birch's pink-brown to white, paper-thin bark peels off in long, horizontal, ribbonlike strips.

HIMALAYAN BIRCH

PAPERBARK MAPLE

### Papery flakes

The Paperbark Maple has a distinctive, flaking bark from a very young age. Its cinnamon-red bark peels in papery flakes to reveal lighter colored fresh bark beneath.

### Patches

As trees age, they may lose patches of old bark because the lenticels become clogged with grime. The London Plane displays a patchwork of light green fresh bark and gray old bark.

LONDON PLANE

# Flowers

Flowers contain a tree's reproductive organs. Hermaphroditic trees, such as cherries, produce flowers with both male and female reproductive organs. Monoecious trees, such as birches, bear separate male and female flowers on the same tree. Dioecious trees, such as hollies, produce male and female flowers on separate trees. How flowers are carried on a tree aids in its identification.

## Single flowers

The flowers on magnolia trees are normally borne individually, with only one flower emerging from each flower bud.

**SAUCER MAGNOLIA**

**SILVER BIRCH**

## Catkins

Catkins are small, often petal-less flowers arranged around a single stem. They may be drooping or upright. Birch trees produce pendulous male catkins and erect female catkins.

## Flower clusters

Flower clusters may be flat-topped, dome-shaped—as with the Common Hawthorne—or upright—as in the case of the Empress Tree's stalked, tubular flowers.

**COMMON HAWTHORN**

**EMPRESS TREE**

# Seeds

Found within nuts, cones, fruit, or berries, seeds are "parcels" that contain all the ingredients needed to produce the next generation of trees. Some may even have wings to aid their dispersal by wind. Seeds are produced from the female flower, or the female part of a flower, once it has been pollinated and fertilized.

**PEACH**

### Acorn

The acorn is the fruit of the oak family. It usually contains a single seed surrounded by a tough, leathery outer shell or casing. Each acorn is held in a rough-textured cup.

**ALGERIAN OAK**

### Edible fruit

Some seeds are wrapped inside brightly colored, sweet-tasting, fleshy fruit. The peach contains just one seed, whereas the apple contains several. These fruit protect the seed until it ripens.

**CORSICAN PINE**

### Cone

Seeds of conifers, such as pines, are carried in a woody, scaly cone. Beneath each scale is a seed attached to a papery wing. These wings enable the seeds to flutter away from the parent tree.

**SYCAMORE**

**SWEET CHESTNUT**

### Winged seed

Some trees, such as the Sycamore, produce winged seeds. The wings help disperse the seeds away from the parent. This ensures that the new tree has reduced competition for natural resources.

### Husk

The seeds of trees such as chestnuts and beeches are encased within a husk, which may be bristly or spiky. Husks split open when the enclosed seeds ripen.

# TREE PROFILES

The trees have first been divided into three main groups: conifers; broadleaves with undivided (simple) leaves; and broadleaves with compound leaves, which have leaves made up of smaller leaflets. Within each of these tree divisions the trees are grouped by leaf shape.

**18** CONIFERS
**40** BROADLEAVES: SIMPLE
**90** BROADLEAVES: COMPOUND

**Symbol**

⭥ Height

**DECIDUOUS**          **EVERGREEN**

Individual artworks show the tree's form, and whether it is evergreen or deciduous.

# 1 CONIFERS

Conifers have existed on Earth for almost 300 million years. They predate flowering trees and bear reproductive organs in cones rather than flowers. They can survive extreme conditions, from freezing, snowy winters to prolonged drought.

**DOUGLAS FIR**
*p.23*

**NOBLE FIR**
*p.21*

## Characteristics

Most conifers develop into large trees, with a single upright stem, rugged bark, and evergreen needlelike leaves. Their seeds are contained in a woody, scaly, conical structure called a cone.

**MONTEREY PINE**
*p.30*

## Location

Adapted to grow at high altitudes, conifers can be found in mountainous regions and in northern temperate zones. Their conical shape enables them to shed snow without branches breaking.

## Forest trees

Most conifers are fast-growing, and are preferred for planting in timber plantations. In shady forest conditions, they grow tall and straight with few low branches.

# Long, thin, single leaves

Trees such as firs and yews fall into this category. Firs bear cones that break up on the tree, while yews have seeds encased in a red berrylike cone, called an aril.

## GRAND FIR

Large American fir, with soft, flattened leaves on either side of the shoots. Leaves are ¾–2 in/ 2–5 cm long; shorter leaves on tip of shoots. When crushed, the foliage smells citrusy. Barrel-shaped cones are up to 4 in/10 cm long.

Slightly notched leaf tip

Leaf dark green on top

Young leaf at shoot tip

Up to 280 ft/80 m

**What to look for** • Leaf shape and arrangement • Shape of leaf tip • Cone shape, orientation, and whether it breaks up on the tree or not

## CAUCASIAN FIR

Densely branched fir, with round-tipped leaves up to 1 in/2.5 cm long—glossy dark green on top, with two white bands on the underside. Unlike other firs shown here, its leaves curve up and away from the shoot.

Male pollen cone below shoot

White bands on leaf underside

✦ Up to 130 ft/40 m

## NOBLE FIR

Leaves have a central groove and a blunt tip. When crushed they emit a slightly "catty" smell. Cylindrical cones, up to 10 in/25 cm long, stand erect on the topmost branches.

Flat leaf up to 1½ in/3.5 cm long

Silver bands on leaf underside

✦ Up to 160 ft/50 m

»

**What to look for** • Leaf color • Leaf shape and arrangement
• Cone shape and size • Bark color and texture

## COMMON YEW

Has dark green and pointed leaves. Olive-green seeds are encased in a red berrylike cone, called an aril. Light brown bark flakes to reveal pink patches. It is often planted in cemetaries.

Leaf up to 1¼ in/3 cm long

Red, fleshy cone

Leaf glossy on top

↕ Up to 70 ft/20 m

## COASTAL REDWOOD

Conical young tree that becomes columnar when mature. Its dark green leaves are carried in two flat rows on either side of the green shoots. Red-brown bark is thick, soft, and fibrous.

Leaf up to ¾ in/2 cm long

Cone ¾–1¼ in/ 2–3 cm long

↕ Up to 360 ft/110 m

LONG, THIN, SINGLE LEAVES | 23

# DOUGLAS FIR

Large, fast-growing
tree, with leaves spirally
arranged on the shoots.
Distinct orange-brown
drooping cones have a
three-pronged bract above each
cone scale. Corky bark becomes vertically
and deeply furrowed when mature.

Leaf
up to
1¼in/
3cm
long

Cone up to
4in/10cm long

Papery bract
above cone scale

↕ Up to 330ft/100m

# WESTERN HEMLOCK

Conical form, with spreading,
arching branches and limp leading
shoots. Soft, pendulous, round-
tipped leaves have two blue-white
bands on the underside. Woody,
egg-shaped cones are up to
1in/2.5cm long.

Dark green leaf up
to ¾in/2cm long

Purplish red
young cone

↕ Up to
280ft/80m

# WOLLEMI PINE

Recently discovered ancient tree; none in
cultivation are older than 20 years. Light
green leaves, borne in rows of
four, turn dark green with age.
Male cones look like clustered
bunches of purple grapes.

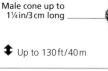

Leaf up to
2in/5cm long

Male cone up to
1¼in/3cm long

↕ Up to 130ft/40m

»

**What to look for** • Leaf shape and arrangement • Fall leaf color • Bark color

## DAWN REDWOOD

Bright green, soft leaves up to ¾in/2 cm long; look fernlike from a distance. They turn gold or plum-colored in the fall before falling. Male and female cones are borne on the same tree. Cinnamon-brown bark peels in vertical flakes.

Downcurved leaf tip

Flattened leaf

↕ Up to 80 ft/25 m

## SWAMP CYPRESS

Conical when young, this tree becomes slightly
domed in maturity. Soft leaves, borne alternately
along the shoots, turn orange-brown in fall.
Unique aerial roots, called "pneumatophores,"
surround the tree base in damp soil. Has dull
reddish brown bark.

Pale green
shoot

Flattened,
bright
green leaf

Up to 150 ft/45 m

# Needlelike, single leaves

The spruces are a group of conifers with sharp-tipped, prickly needles. They have pendulous cones, which do not break up on the tree but drop intact.

**What to look for** • Tree form • Leaf shape and arrangement • Cone shape, texture, and orientation • Bark texture

## SITKA SPRUCE

Easily identified by its rigid, sharply pointed leaves, which are radially arranged around pale brown shoots. Pendulous, pale brown cones are up to 4 in/10 cm long. Its purple-brown bark lightens and develops cracks with age.

Blue-green leaf

Leaf up to 1¼ in/3 cm long

Thin, papery cone scale

Up to 315 ft/95 m

## BREWER'S WEEPING SPRUCE

Ornamental conifer, with long, pendulous branchlets up to 6 ft/2 m in length, covered in flexible needles up to 1½ in/3.5 cm long. Has chestnut-brown, pendulous cones. Its gray-purple bark cracks into irregular plates.

Black-green leaf

Leaf tapers to fine point

Up to 180 ft/55 m

Cone up to 5 in/12 cm long

## SERBIAN SPRUCE

Slender, spirelike form. Branches sweep downward, but arch upward at the tips. Glossy green leaves have two white bands on the underside. Teardrop-shaped cones are purple-brown. Orange-brown to copper bark sheds in irregular plates.

Blunt
leaf tip

Stiff leaf up to
¾ in/2 cm long

✤ Up to 160 ft/50 m

## ORIENTAL SPRUCE

Glossy,
rigid leaf

Conical, densely branched conifer. Short, blunt-tipped leaves are shiny dark green on top, pale green on the underside. Pendulous, cylindrical cones, up to 4 in/10 cm long, are purple ripening to brown. Pinkish brown bark flakes into irregular plates.

Leaf up to
½ in/1 cm long

✤ Up to
180 ft/55 m

## NORWAY SPRUCE

Traditional Christmas tree for most of Europe. Stiff, needlelike leaves up to ¾ in/2 cm long are borne all around tobacco-brown shoots. Pendulous, banana-shaped cones are up to 6 in/ 15 cm long. Coppery pink, smooth bark develops shallow plates.

Green cone
turns brown
with age

Sharply
pointed
leaf

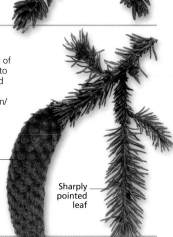

✤ Up to
160 ft/50 m

# Cones

Unique to conifers, cones are an important clue to identification of these trees. Most cones are hard and woody, but some may be fleshy and edible.

Conifers bear separate male and female cones on the same tree. Male cones produce pollen. Female cones produce ovules and, once fertilized, the seeds. Some cones can take several years to ripen, open, and disperse seeds. Edible cones are eaten by animals and birds, who then disperse the seeds by excreting them.

## Pollination

Wind pollination is a hit-and-miss affair, so conifers such as the Scots Pine produce huge quantities of pollen. Male cones ripen to release this pollen into the air for wind dispersal, sometimes creating a yellow cloud around the tree. Some pollen falls on the female cones, fertilizing the ovules on their scales.

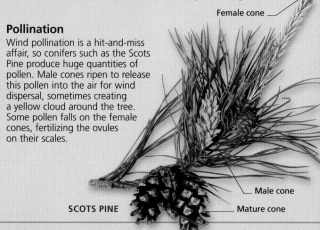

Female cone

Male cone

Mature cone

**SCOTS PINE**

## Ripe cones

Once the ovules are fertilized, seeds are formed and the cones ripen from green to woody brown. When ripe, the Incense Cedar's six-scaled cones break up on the tree. Monterey Pine cones take several years to open and release seeds. The open cones may remain on the tree for several years after seed dispersal.

Bulging cone scale

Red-brown cone

Open, woody brown cone

**INCENSE CEDAR**

**MONTEREY PINE**

## Cone orientation

Trees such as the Atlas Cedar bear cones that are held erect on the top of branches. These cones ripen and then break up in position, releasing their seeds in the process. Other trees, such as the Bhutan Pine, have cones that hang beneath the branches, ripen, disperse their seeds, and then fall without breaking apart.

Brown
erect cone

Pale brown
drooping cone

**BHUTAN PINE**                    **ATLAS CEDAR**

## Cone size

Cone size varies greatly. The Lawson Cypress bears small globular cones, less than ½ in (1 cm) across in profusion along the branch tips. At the other end of the scale, the Mexican White Pine's cones are big and drooping. Borne on stout stalks, they can grow up to 18 in (45 cm) long.

Large
cone scale

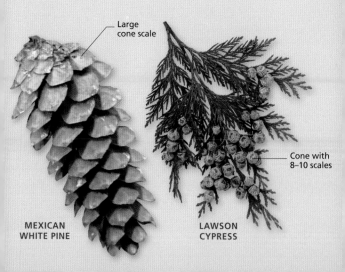

Cone with
8–10 scales

**MEXICAN**
**WHITE PINE**

**LAWSON**
**CYPRESS**

# Needlelike leaves in clusters

This group of trees includes pines and cedars. Pines bear leaves in pairs, threes, or fives, while cedar leaves are borne in dense whorls.

**What to look for** • Leaf shape and arrangement • Pink sheath at leaf base • Cone texture and orientation

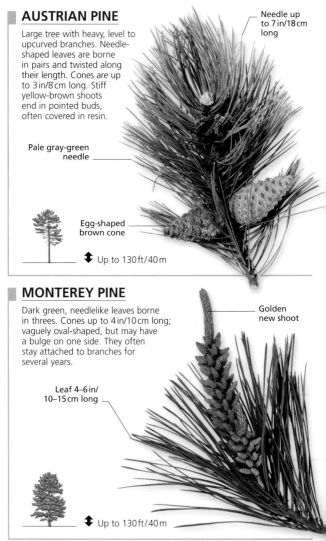

## AUSTRIAN PINE

Large tree with heavy, level to upcurved branches. Needle-shaped leaves are borne in pairs and twisted along their length. Cones are up to 3 in/8 cm long. Stiff yellow-brown shoots end in pointed buds, often covered in resin.

Needle up to 7 in/18 cm long

Pale gray-green needle

Egg-shaped brown cone

‡ Up to 130 ft/40 m

## MONTEREY PINE

Dark green, needlelike leaves borne in threes. Cones up to 4 in/10 cm long; vaguely oval-shaped, but may have a bulge on one side. They often stay attached to branches for several years.

Golden new shoot

Leaf 4–6 in/ 10–15 cm long

‡ Up to 130 ft/40 m

## SCOTS PINE

Most common pine in Europe. Blue-green leaves, up to 3 in/7 cm long, are borne in pairs and twisted along their length. Male and female cones are borne on young shoots on the same tree. Its distinctive, orange-red flaking bark is most obvious on older trees.

Female cone

Male cone

Egg-shaped, woody, mature seed cone

↕ Up to 100 ft/30 m

## BHUTAN PINE

Decorative pine with soft, drooping leaves, borne in clusters of five. Banana-shaped cones, up to 12 in/30 cm long, often have scales covered with white, crystallized resin.

Pollen-bearing male cone

Leaf up to 8 in/20 cm long

↕ Up to 130 ft/40 m

## WESTERN YELLOW PINE

Its gray-green leaves, held in clusters of three, face forward on the shoot. Red-brown egg-shaped cones are up to 4 in/10 cm long, with a spine on each scale.

Leaf up to 10 in/25 cm long

Purple male cone

↕ Up to 160 ft/50 m

»

**What to look for** • Tree form • Leaf shape and arrangement
• Cone shape, size, and orientation

## ATLAS CEDAR

Wide-spreading tree when mature,
with upcurved branch tips. Blue-green
leaves are borne in dense whorls on
side shoots and singly on leading shoots.
Cones, up to 3 in/7.5 cm long, are borne
on top of branches.

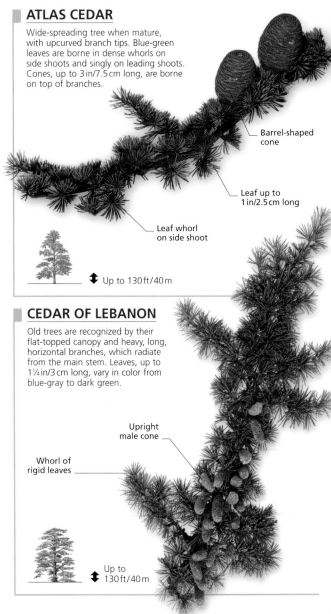

Barrel-shaped
cone

Leaf up to
1 in/2.5 cm long

Leaf whorl
on side shoot

↕ Up to 130 ft/40 m

## CEDAR OF LEBANON

Old trees are recognized by their
flat-topped canopy and heavy, long,
horizontal branches, which radiate
from the main stem. Leaves, up to
1¼ in/3 cm long, vary in color from
blue-gray to dark green.

Upright
male cone

Whorl of
rigid leaves

↕ Up to
130 ft/40 m

## DEODAR CEDAR

Distinguished from other cedars by its
drooping branch tips. Leaves are borne
in whorls on side shoots and singly on
leading shoots. Blue-green when young,
they turn dark green with age.

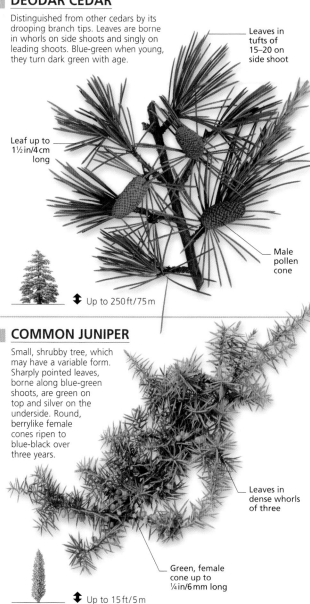

Leaves in
tufts of
15–20 on
side shoot

Leaf up to
1½in/4cm
long

Male
pollen
cone

‡ Up to 250ft/75m

## COMMON JUNIPER

Small, shrubby tree, which
may have a variable form.
Sharply pointed leaves,
borne along blue-green
shoots, are green
on top and silver on the
underside. Round,
berrylike female
cones ripen to
blue-black over
three years.

Leaves in
dense whorls
of three

Green, female
cone up to
¼in/6mm long

‡ Up to 15ft/5m

»

**What to look for** • Tree form • Leaf shape and arrangement • Pink sheath at leaf base • Fall leaf color

# EUROPEAN LARCH

Broadly conical tree, distinguished from other larches by its pendulous, straw-colored shoots and egg-shaped cones. Leaves are bright green when young, turning golden yellow in fall. Cones, up to 1¾ in/4.5 cm long, have inward-curving scales.

Leaf up to 1½ in/4 cm long

Blunt, upright cone

Up to 130 ft/40 m

# JAPANESE LARCH

Distinguished from the European Larch by its reddish purple shoots—conspicuous on leafless branches in winter. Scales on the squat, bun-shaped cones turn outward and downward.

Cone up to 1¼ in/ 3 cm long and wide

Leaf ¾ in/2 cm long

Up to 120 ft/35 m

# HYBRID LARCH

Broadly conical tree that grows faster than the European or Japanese Larch. Leaves borne singly on leading shoots and in dense rosettes elsewhere. Orange-brown shoots not as pendulous as on the European Larch.

Leaf up to 3 in/8 cm long

Cone scale bends outward, not downward

Up to 110 ft/32 m

# Fan-shaped leaves

This tree is the only surviving species of those primitive trees, which had fan-shaped leaves. It evolved more than 200 million years ago, before most other conifers.

**What to look for** • Tree form • Leaf shape and arrangement
• Fall leaf color

## MAIDENHAIR TREE

Upright, broadly conical tree. Unique leaves are borne on long, slender stalks, and turn golden yellow in fall. Yellow-green seed, present only on female trees, contains kernel. Pale gray-brown bark cracks with age. Also known as the Ginkgo.

Deeply notched leaf

Leaf up to 3 in/8 cm wide

Veins radiate from leaf base

Fleshy seed up to 1½ in/4 cm long

Up to 130 ft/40 m

# Scalelike leaves

The leaves on these trees are either borne in flattened sprays or pressed close to the shoot. The Monkey Puzzle and Japanese Cedar have triangular, scalelike leaves.

**What to look for** • Leaf shape and arrangement • Leaf scent • Cone shape and orientation

## ITALIC CYPRESS

One of the most easily recognizable conifers, with a distinctive tight, spirelike form. Leaves emit little noticeable scent when crushed. Upcurved branches obscure the trunk from view.

Lumpy, glossy green cone

Dark, dull green leaf

Red-brown shoot

⬍ Up to 60 ft/18 m

## LAWSON CYPRESS

Fast-growing tree, with dark green sprays of soft, scalelike leaves. They smell of parsley when crushed. Green-purple seed cones, ¼in/7mm across with 8–10 scales, ripen to slate-blue, then woody brown.

Small, pointed leaf

Globular seed cone

↕ Up to 130ft/40m

## MONTEREY CYPRESS

Common in coastal regions, this tree is easily recognized by its flat-topped, spreading crown of large, heavy branches, similar to the Cedar of Lebanon (p.32). Mature cones are globular and up to 1¼in/3.5cm across.

Leaf has pointed tip

Green, rosettelike young cone

↕ Up to 130ft/40m

## LEYLAND CYPRESS

Fast-growing, cultivated hybrid, commonly planted as hedges and screens. Light green leaves, irregularly arranged on red-brown shoots, emit a pungent scent when crushed. Male and female cones are borne on the same tree.

Small leaf has pointed tip

Flattened leaves borne in sprays

↕ Up to 122ft/36m

»

**What to look for** • Tree form • Leaf shape and arrangement • Leaf tip • Type of branching • Bark color and texture

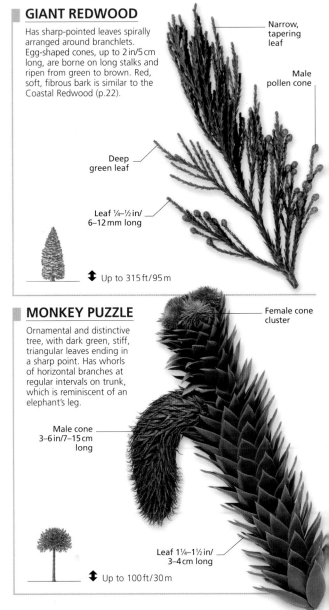

## GIANT REDWOOD

Has sharp-pointed leaves spirally arranged around branchlets. Egg-shaped cones, up to 2 in/5 cm long, are borne on long stalks and ripen from green to brown. Red, soft, fibrous bark is similar to the Coastal Redwood (p.22).

Narrow, tapering leaf

Male pollen cone

Deep green leaf

Leaf ¼–½ in/ 6–12 mm long

⬍ Up to 315 ft/95 m

## MONKEY PUZZLE

Ornamental and distinctive tree, with dark green, stiff, triangular leaves ending in a sharp point. Has whorls of horizontal branches at regular intervals on trunk, which is reminiscent of an elephant's leg.

Female cone cluster

Male cone 3–6 in/7–15 cm long

Leaf 1¼–1½ in/ 3–4 cm long

⬍ Up to 100 ft/30 m

## JAPANESE CEDAR

Fast-growing conifer with downward-curving branches. Its soft, red-brown bark peels vertically. Bright green shoots bear spiraling, stiff leaves. Broad-based leaves taper to a soft point.

Stiff, upright stalk

Cone up to ¾ in/2 cm wide

Leaf up to ½ in/1.5 cm long

 Up to 200 ft/60 m

## WESTERN RED CEDAR

Has sprays of scalelike foliage similar to the Lawson Cypress (p.37), but wider and flatter. When crushed, leaves smell like ripe pineapple. Fibrous, red-brown bark peels in long, thin strips.

Thick, blunt leaf

Leaf bright green on top

 Up to 290 ft/85 m

## INCENSE CEDAR

Conical, columnlike tree with upcurved branches. Dense foliage is carried in flat sprays of overlapping scales. Crushed foliage smells like shoe polish. Reddish brown bark becomes cracked and flaky with age.

Mid- to dark green foliage

Cone scales open wide

Leaf up to ⅛ in/3 mm long

 Up to 160 ft/50 m

# 2 BROADLEAVES: SIMPLE

As the name suggests, broadleaved trees have broad leaves rather than needles or scales. A large number of broadleaved trees have simple, undivided leaves. They may be evergreen or deciduous, and produce a variety of leaf shapes.

**WILLOW-LEAVED PEAR**
*p.64*

## Flowers, fruit, and seed

After flowering, a broadleaf tree produces fruit. These fruit may be in various forms, from the winged seed of maples to the fleshy fruit of plums and pears.

**RED MAPLE**
*p.43*

**PEDUNCULATE OAK**
*p.47*

**NORWAY MAPLE**
*p.42*

**PLUM**
*p.78*

## Leaf shapes

There is wide variety in leaf shapes for broadleaves. Some trees have hand-shaped leaves, such as maples, and others have lobed leaves, such as oaks.

# Hand-shaped leaves

Trees with palmately lobed, or hand-shaped, leaves include maples and planes. Most maples have good leaf color in the fall, and planes have a distinctive bark.

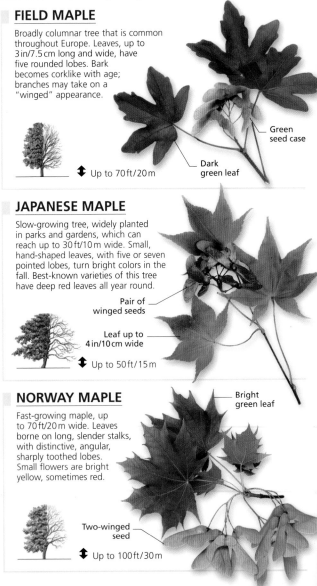

## FIELD MAPLE

Broadly columnar tree that is common throughout Europe. Leaves, up to 3 in/7.5 cm long and wide, have five rounded lobes. Bark becomes corklike with age; branches may take on a "winged" appearance.

Green seed case

Dark green leaf

↕ Up to 70 ft/20 m

## JAPANESE MAPLE

Slow-growing tree, widely planted in parks and gardens, which can reach up to 30 ft/10 m wide. Small, hand-shaped leaves, with five or seven pointed lobes, turn bright colors in the fall. Best-known varieties of this tree have deep red leaves all year round.

Pair of winged seeds

Leaf up to 4 in/10 cm wide

↕ Up to 50 ft/15 m

## NORWAY MAPLE

Bright green leaf

Fast-growing maple, up to 70 ft/20 m wide. Leaves borne on long, slender stalks, with distinctive, angular, sharply toothed lobes. Small flowers are bright yellow, sometimes red.

Two-winged seed

↕ Up to 100 ft/30 m

**What to look for** • Leaf size, color, and margin • Number and shape of leaf lobes • Fruit (winged seed) color

## RED MAPLE

Broadly columnar maple, with three or five shallow lobes to each leaf. Leaves, dark matte green on top and blue-green on the underside, turn scarlet in the fall.

Coarsely serrated leaf margin

Leaf up to 4 in/10 cm long and wide

↕ Up to 100 ft/30 m

## SUGAR MAPLE

Broadly columnar maple native to North America; produces maple syrup. Leaves, pale green on the underside, have some hairs on vein axils, and turn vibrant yellow, orange, and red in the fall.

Broad leaf lobe tapers to long point

Five-lobed leaf

Leaf mid- to dark green on top

↕ Up to 120 ft/35 m

## SYCAMORE

Fast-growing maple, up to 70 ft/20 m wide. Its large leaves are borne on pink-green stalks. In spring, clusters of yellow, drooping flowers appear.

Deep green leaf

Leaf up to 8 in/20 cm wide

↕ Up to 100 ft/30 m

»

**What to look for** • Tree form • Leaf shape, color, and arrangement • Fruit color and shape • Bark color and texture

## SWEETGUM

Often confused with maples due to its leaf shape. This tree's leaves are alternately positioned on corky shoots, whereas maple leaves are oppositely arranged on smooth shoots. Fruit are like those of plane trees.

Long leaf stalk

Tapering leaf lobe

Glossy green leaf

Up to 150 ft/45 m

## WHITE POPLAR

Broadly conical, this tree's distinctive variously lobed leaves are densely covered with hairs on the underside. Green capsulelike fruit is covered in fluffy, white hairs. Light gray bark is pitted with black, diamond-shaped patches.

Light green leaf veins

Leaf shiny dark green on top

Up to 84 ft/26 m

White hairs on leaf underside

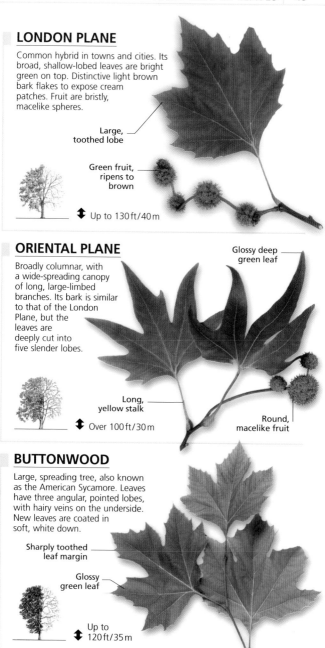

# LONDON PLANE

Common hybrid in towns and cities. Its broad, shallow-lobed leaves are bright green on top. Distinctive light brown bark flakes to expose cream patches. Fruit are bristly, macelike spheres.

Large, toothed lobe

Green fruit, ripens to brown

Up to 130 ft/40 m

# ORIENTAL PLANE

Broadly columnar, with a wide-spreading canopy of long, large-limbed branches. Its bark is similar to that of the London Plane, but the leaves are deeply cut into five slender lobes.

Glossy deep green leaf

Long, yellow stalk

Over 100 ft/30 m

Round, macelike fruit

# BUTTONWOOD

Large, spreading tree, also known as the American Sycamore. Leaves have three angular, pointed lobes, with hairy veins on the underside. New leaves are coated in soft, white down.

Sharply toothed leaf margin

Glossy green leaf

Up to 120 ft/35 m

# Lobed leaves

Several trees, including most oaks, bear leaves that are lobed to a greater or lesser degree. All oaks produce seeds called acorns.

**What to look for** • Tree form • Leaf shape and arrangement • Fruit shape and size

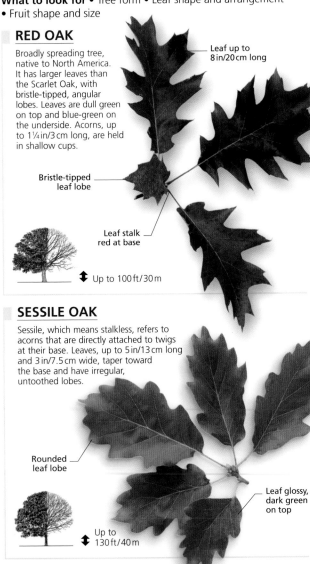

## RED OAK

Broadly spreading tree, native to North America. It has larger leaves than the Scarlet Oak, with bristle-tipped, angular lobes. Leaves are dull green on top and blue-green on the underside. Acorns, up to 1¼ in/3 cm long, are held in shallow cups.

Leaf up to 8 in/20 cm long

Bristle-tipped leaf lobe

Leaf stalk red at base

↕ Up to 100 ft/30 m

## SESSILE OAK

Sessile, which means stalkless, refers to acorns that are directly attached to twigs at their base. Leaves, up to 5 in/13 cm long and 3 in/7.5 cm wide, taper toward the base and have irregular, untoothed lobes.

Rounded leaf lobe

Leaf glossy, dark green on top

↕ Up to 130 ft/40 m

## PEDUNCULATE OAK

Also known as the English Oak. Distinguished from the Sessile Oak by slightly bigger acorns, which are borne on stalks up to 4 in/10 cm long. Its elliptic leaves have rounded, irregular lobes, which reduce in size toward the base.

Leaf deep green on top

Acorn up to 1½ in/4 cm long

 Up to 130 ft/40 m

## SCARLET OAK

The leaves of this tree has leaves that turn vibrant scarlet in the fall. Three angular lobes on each side of the leaf cut it almost to the midrib. Acorns, up to ¾ in/2 cm long, are held in a shallow, scaly cup.

Leaf up to 6 in/15 cm long

Angular leaf lobe

Up to 100 ft/30 m

## TURKEY OAK

Wide-spreading tree, with glossy, dark green leaves. Variably lobed, with shallower lobes than either the Sessile or Pedunculate Oak. Acorns up to 1 in/2.5 cm long are held in a bristly cup.

Bristly acorn cup

Leaf up to 5 in/ 12 cm long

Up to 130 ft/40 m

**What to look for** • Leaf shape and lobes • Fruit color, shape, and size

## HAWTHORN

Small, hardy tree with thorns; often found in hedgerows. Its variably shaped leaves are followed by pungent-smelling flower clusters in spring. Berrylike fruit, called haws, ripen from green to red.

Leaf deeply cut toward midrib

Creamy white flower

Up to 40 ft/12 m

## MIDLAND HAWTHORN

Similar in size and shape to the Hawthorn but with more shallowly lobed leaves. Glossy, dark green leaves have toothed margins. Red berrylike fruit, up to 1 in/2.5 cm long, are larger than those of the Hawthorn.

Toothed leaf margin

Oval, red fruit

Leaf up to 2 in/5 cm long

Up to 30 ft/10 m

## TULIP TREE

This tree has unusually shaped leaves, with four angular lobes and cut-off, indented tips. Leaves are up to 6 in/15 cm long and 4 in/10 cm wide. Tulip-shaped flowers are yellow-green with an orange flush.

Butter-yellow leaf in fall

Orange-yellow stamen

Up to 160 ft/50 m

Cone-shaped fruit

# Inversely egg-shaped leaves

Trees with inversely egg-shaped, or obovate, leaves include the Common Alder and the Persian Ironwood. These leaves are broadest above the middle.

**What to look for** • Leaf shape and size • Flower color and arrangement

## COMMON ALDER

Broadly conical, its leaves are shiny green on top, gray-green on the underside, with tufts of hair in the vein axils. It bears catkins and its fruit, shaped like small green pineapples, ripen to brown woody cones.

Unripe fruit up to ¾ in/2 cm long

Leaf up to 4 in/10 cm long

↕ Up to 80 ft/25 m

## PERSIAN IRONWOOD

Broadly spreading, large tree with flaking bark like the London Plane (p.45). Leaves, up to 2½ in/6 cm wide, broadest toward their tip, turn brilliant red and gold in the fall. Small, red flowers appear on bare branches in winter.

Leaf up to 5 in/12 cm long

Fall leaf

Shallowly toothed leaf margin

↕ Up to 70 ft/20 m

# Egg-shaped leaves

A wide variety of trees bear egg-shaped, or ovate, leaves. While some leaves have a heart-shaped base, others have a strongly serrated margin.

### DOWNY BIRCH

Broadly conical tree, which differs from the Silver Birch (p.85) in its soft, hairy shoots and less pendulous branches. Dark green leaves have a paler underside and hairy veins.

Leaf up to 2in/5cm long and wide

Toothed leaf margin

‡ Up to 80ft/25m

### CHINESE RED-BARKED BIRCH

Large, conical birch native to Asia. Can be recognized by its coppery, orange-red bark, which peels to reveal cream-pink patches. Glossy mid-green leaves, up to 3in/7.5cm long, turn golden yellow in fall.

Drooping male catkin

Serrated leaf margin

BARK

‡ Up to 80ft/25m

### HIMALAYAN BIRCH

Broadly conical birch with red-brown to white bark. Serrated leaves, up to 4in/10cm long, are distinctly lined with up to 12 pairs of parallel veins.

Vein ends in sharp tooth

Pale green leaf underside

Upright, green female catkin

‡ Up to 80ft/25m

**What to look for** • Tree form • Leaf shape, base, and margin

## GRAY ALDER

Broadly conical tree with smooth, dark gray bark. Leaves are matte green on top, with gray hairs on the underside. Flowers borne in long catkins in early spring; male catkins are yellow and pendulous.

Leaf up to 4 in/10 cm long

Conelike fruit

Irregular leaf margin

✦ Up to 70 ft/20 m

## ITALIAN ALDER

Large, broadly conical tree. Leaves, up to 4 in/10 cm long, are bright, glossy green on top, with a sparsely hairy underside. Leaf base is heart-shaped. Smooth, dull gray bark cracks with age.

Unripe fruit

Pale green leaf underside

Toothed leaf margin

✦ Up to 100 ft/30 m

## SNOWY MESPIL

Small, bushy, multi-stemmed tree. Flowers appear in upright clusters as leaves unfurl from winter buds. Bright, coppery pink leaves fade to yellow-green. Bears round, black-purple berries.

Pointed leaf tip

Starlike white flower

✦ Up to 40 ft/12 m

»

**What to look for** • Leaf shape, size, and margin • Flower color and size • Fruit shape and size

# HACKBERRY

Leaves resemble those of the Common Stinging Nettle: saw-toothed around the margin (except for lower third of leaf). Rounded fruit, borne singly on thin green stalks, turn from green to red.

Fruit up to ½ in/1 cm across

Leaf up to 5 in/12 cm long, 2 in/5 cm wide

↕ Up to 80 ft/25 m

# COMMON HORNBEAM

Serrated leaf margin

Male catkin

Similar to the Common Beech (p.55), except for its 10 to 13 pairs of parallel leaf veins, three-lobed papery bracts around the seeds, and angular trunk and branches. Male and female catkins are borne on the same tree.

Female catkin at shoot tip

↕ Up to 100 ft/30 m

# INDIAN BEAN TREE

Leaf broadly egg-shaped

Wide-spreading, open-branched tree. Large, soft, limp leaves, up to 10 in/25 cm long and 6 in/15 cm wide, are grass-green on top. Upright clusters of orchidlike flowers appear in mid- to late summer, followed by seed pods up to 12 in/30 cm long.

Pink-white flower

Long, thin seed pod

Up to ↕ 70 ft/20 m

## PAPER-BARK BIRCH

Distinctive creamy white
bark peels in thin, papery
strips. Dark green
leaves are smooth on
top. Paler leaf underside
has veins lightly covered
in fine hairs. Pendulous
yellow catkins appear
in spring.

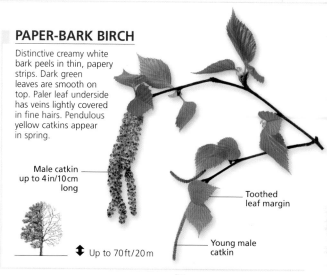

Male catkin
up to 4 in/10 cm
long

Toothed
leaf margin

Young male
catkin

↕ Up to 70 ft/20 m

## TURKISH HAZEL

Has a distinctive pyramid-shaped
canopy on a short, straight trunk.
Leaves are broadly egg-shaped, with
an irregularly serrated margin and
heart-shaped base. Fruit (hazelnut)
borne in husk. Gray-brown bark
turns corklike with age.

Green,
bristly husk

Leaf up to
4 in/10 cm wide

↕ Up to 80 ft/25 m

## POCKET HANDKERCHIEF TREE

Easily recognized in spring by large flower
bracts, which hang from branches like
handkerchiefs. Egg-shaped
leaves have a pointed tip
and heart-shaped base.
Green, circular husk
ripens to purple-brown.

Leaf glossy
green on top

Limp, white bract up
to 8 in/20 cm long

↕ Up to 80 ft/25 m

»

**What to look for** • Tree form • Leaf shape, size, and margin • Leaf texture • Fruit shape and size • Bark color

## KEYAKI

Broadly spreading tree, with smooth, gray bark similar to that of the Common Beech. Rough leaves, on red-brown zigzag twigs, turn orange-red in the fall, before falling. Bears very small winter buds.

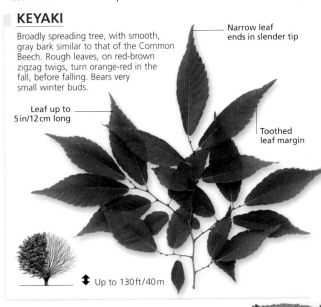

Narrow leaf ends in slender tip

Leaf up to 5 in/12 cm long

Toothed leaf margin

⬍ Up to 130 ft/40 m

## ENGLISH ELM

Broadly columnar tree, distinguished by its rough leaves, which have a pointed tip and asymmetrical base. Seeds are contained within clusters of oval-shaped, papery wings. Bark is gray-brown.

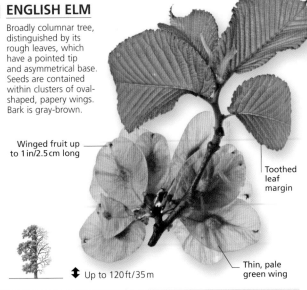

Winged fruit up to 1 in/2.5 cm long

Toothed leaf margin

Thin, pale green wing

⬍ Up to 120 ft/35 m

## COMMON BEECH

Broadly spreading tree with a smooth, silver-gray bark. Leaves are wavy-edged, but untoothed. Husk, covered in coarse bristles, contains up to three triangular seeds. It ripens from green to brown.

Woody husk

Leaf up to 4 in/10 cm long

↕ Up to 130 ft/40 m

## PURPLE BEECH

Almost identical to the Common Beech, except for its slightly smaller leaves, ranging in color from purple-green to deep purple, almost black. Leaf color is most vibrant in spring.

Leaf up to 3½ in/9 cm long

Seed husk, covered in coarse bristles

↕ Up to 130 ft/40 m

## ORIENTAL BEECH

More vigorous and potentially larger than the Common Beech. Its leaves are bigger, with up to 12 pairs of veins, rather than eight. Smooth, light gray bark sometimes has shallow cracks in maturity.

Leaf up to 5 in/12 cm long

Wavy but untoothed leaf margin

↕ Up to 100 ft/30 m

»

**What to look for** • Leaf shape, size, and margin • Flower color and size • Fruit shape and size

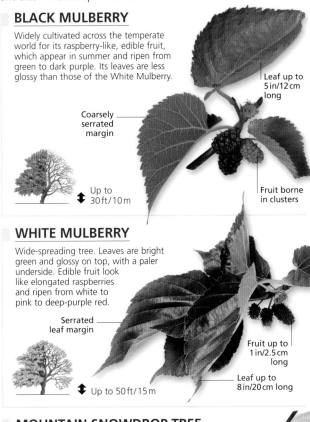

# BLACK MULBERRY

Widely cultivated across the temperate world for its raspberry-like, edible fruit, which appear in summer and ripen from green to dark purple. Its leaves are less glossy than those of the White Mulberry.

Leaf up to 5 in/12 cm long

Coarsely serrated margin

Up to 30 ft/10 m

Fruit borne in clusters

# WHITE MULBERRY

Wide-spreading tree. Leaves are bright green and glossy on top, with a paler underside. Edible fruit look like elongated raspberries and ripen from white to pink to deep-purple red.

Serrated leaf margin

Up to 50 ft/15 m

Fruit up to 1 in/2.5 cm long

Leaf up to 8 in/20 cm long

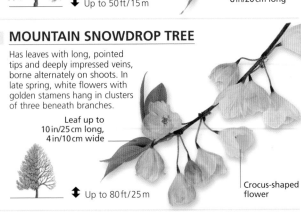

# MOUNTAIN SNOWDROP TREE

Has leaves with long, pointed tips and deeply impressed veins, borne alternately on shoots. In late spring, white flowers with golden stamens hang in clusters of three beneath branches.

Leaf up to 10 in/25 cm long, 4 in/10 cm wide

Up to 80 ft/25 m

Crocus-shaped flower

## COMMON CRAB APPLE

Broadly spreading tree with twisting branches. In spring, apple-blossom-like flowers appear. They are white with a pink blush. Small, bitter-tasting, fruit appear in late summer; these are green, sometimes flushed with red.

Leaf up to 1½in/4cm long

Applelike fruit, up to 1½in/4cm across

⬍ Up to 30ft/10m

## SIBERIAN CRAB APPLE

Widely grown, medium-sized, ornamental tree. Has attractive, fragrant flowers, with white petals surrounding a cluster of yellow anthers. Rounded fruit, yellow-green ripening to red, are borne singly on a stiff stalk.

Leaf up to 3in/8cm long

Fruit up to ¾in/2cm across

⬍ Up to 50ft/15m

## CULTIVATED APPLE

Small, round-headed tree, grown for its edible fruit. Leaves are deep green on top, with serrated margins. Fragrant, five-petaled flowers are white, tinged with pink that fades over time.

Leaf up to 5in/12cm long

Fruit up to 3½in/9cm across

Up to 30ft/10m

Pale green leaf underside

»

**What to look for** • Tree form • Leaf shape, size, and margin
• Leaf color and texture

## ANTARCTIC BEECH

Broadly columnar tree, with
small leaves up to 1½ in/3 cm
long and ¾ in/2 cm wide.
They are crinkled and
slightly cupped around
the margin. These glossy
leaves may look
evergreen but
they are not.

Irregularly toothed
leaf margin

Leaf deep green
on top

Fruit
husks in
cluster

↕ Up to 50 ft/15 m

## ROBLE BEECH

Large South American tree
with egg-shaped to oval,
occasionally oblong, leaves.
Dark green on top
and blue-green on
the underside, they
have 8 to 11 pairs
of distinct veins.

Leaf up to
3 in/7.5 cm long

Green-brown
husk

Irregularly
toothed
leaf margin

↕ Up to 120 ft/35 m

## CHINESE NECKLACE POPLAR

Distinctive wide-spreading tree. Has an open, gangly form and long, stiff branches. Large leaves have a heart-shaped base. Long, pendulous catkins release large amounts of cotton-wool-like seeds in summer.

Leaf bright green on top

Leaf up to 14 in/35 cm long, 8 in/20 cm wide

↕ Up to 70 ft/20 m

## APRICOT

Densely branched, broadly spreading tree. Leaves have a rounded base, pointed tip, and serrated margin. Pale pink to white five-petaled flowers, up to 1½ in/4 cm across. Peachlike yellow-orange fruit, up to 1½ in/3.5 cm across.

Leaf up to 3½ in/9 cm long, 3 in/7 cm wide

↕ Up to 40 ft/12 m

## EMPRESS TREE

Flower up to 2 in/5 cm long

Fast-growing, wide-spreading ornamental tree. Large, limp, soft leaves are dark green on top and are covered in fine hairs. Widely grown for its beautiful lilac-purple, trumpet-shaped flowers borne in erect clusters.

Leaf up to 14 in/30 cm long and wide

Light green leaf underside

↕ Up to 70 ft/20 m

»

**What to look for** • Tree form • Leaf shape, size, and margin • Leaf color and texture

## COIGÜE SOUTHERN BEECH

Broadly conical tree, with small leaves so densely borne that the tree seems black in color. Glossy leaves are dark green on top and have a matte, pale green underside.

Leaf 1–1½ in/ 2.5–4 cm long

Serrated leaf margin

Up to 77 ft/23 m

## CORK OAK

Native to the Mediterranean region, its thick, soft bark is used to make cork products, such as wine stoppers. Dark green leaves, paler on the underside, have broadly toothed margins.

Shiny leaf

Leaf up to 3 in/7.5 cm long

Up to 70 ft/20 m

## PORTUGUESE LAUREL

Shrubby tree, with creamy white flowers held in long, narrow clusters. Multiple stems have green leaves with shallow, roundly serrated margin.

Glossy leaf up to 4 in/10 cm long, 2 in/5 cm wide

Flower cluster up to 10 in/25 cm long

Up to 30 ft/10 m

# Fernlike leaves

The very first trees reproduced from spores, tree ferns, are the only spore-producing trees today. They have long, deeply cut leaves (fronds) and soft, fibrous bark.

**What to look for** • Tree form • Leaf shape, size, and color
• Leaf tip • Bark color and texture

### SOFT TREE FERN

Hardy tree fern from Australasia. Leaf fronds, up to 12 ft/4 m long, regrow each year from the top of the cylindrical trunk. Its chestnut-brown bark is soft and fibrous.

Tiny, pointed leaflet

Leaflet shiny, rich green on top

⬍ Up to 22 ft/7 m

# Lance-shaped leaves

Trees such as willows have lance-shaped, or lanceolate, leaves about five times as long as they are broad. They taper toward each end, although not always equally.

**What to look for** • Tree habit • Leaf shape, size, and color • Shoot color and orientation

## CRACK WILLOW

Up to 70 ft/20 m wide, fast-growing tree found alongside rivers and streams. Takes its name from the distinctive crack heard when its twigs snap. Leaves are up to 6 cm/15 cm long and 1 in/2.5 cm wide.

Raised central midrib

Leaf glossy, dark green on top

✥ Up to 80 ft/25 m

## GOLDEN WEEPING WILLOW

Distinctive hybrid, widely grown alongside rivers, lakes, and ponds for its ornamental golden shoots and graceful drooping habit. Leaves are up to 5 in/13 cm long and 1 in/2.5 cm wide.

Leaf tapers at end

Golden yellow shoot

Bright yellow catkin

✥ Up to 70 ft/20 m

## COMMON OSIER

Shrubby tree, with long, upright stems grown for basket-making canes. Smaller than the Golden Weeping Willow and the Crack Willow, its long, slender leaves are gray-haired on the underside.

Leaf dark green on top

✥ Up to 25 ft/8 m

Leaf 4–10 in/10–25 cm long, ¼–½ in/0.5–2 cm wide

# Lance-shaped to elliptic leaves

These eucalyptus trees from the Southern Hemisphere have leaves that can be lance-shaped or narrowly elliptic—broadest at the center, with tapering ends.

**What to look for** • Leaf color, shape, and arrangement
• Bark color

## TASMANIAN BLUE GUM

Large tree, with leathery leaves like curving spears. Leaves are up to 12 in/30 cm long and 2 in/5 cm wide. Its gray-brown bark falls in strips to reveal smooth white or pink-white bark beneath.

Leaf deep blue-green on top

Up to 130 ft/40 m

## CIDER GUM

Broadly columnar, fast-growing tree. Silver-blue juvenile leaves, leathery and stalkless, are borne in opposite pairs. These become spear-shaped when mature, and are borne on light green stalks.

Blue-green mature leaf

Oval, woody seed capsule

Leaf up to 5 in/ 12 cm long, 1½ in/4 cm wide

Up to 100 ft/30 m

»

**What to look for** • Tree form • Leaf shape, and arrangement • Leaf color • Flower arrangement and color

## PEACH

Small, spreading tree cultivated for its edible, round, fleshy fruit, which contains a stonelike seed. Leaves are deep green on top, paler on the underside. Cherry blossom-like flowers appear in early spring before the leaves.

Leaf up to
6 in/16 cm long

Pale to deep
pink flower

Up to
30 ft/10 m

Flower
borne on
short stalk

## TIBETAN CHERRY

Broadly spreading tree, recognized by its smooth, red-brown bark, which peels in thin, horizontal strips with age. Produces small, white, relatively inconspicuous flowers in spring.

Egg-shaped,
crimson
berry

Toothed
leaf with
pointed tip

Up to 50 ft/15 m

**BARK**

## WILLOW-LEAVED PEAR

Five-petaled
flower

Small, weeping tree, also known as the Silver-leaved Pear. Gray-green leaves twist along their length and are covered in silver-white hairs on the underside. Fruit is a small hard pear, up to 1¼ in/3 cm long.

Lance-shaped
leaf up to
4 in/10 cm long

Up to 30 ft/10 m

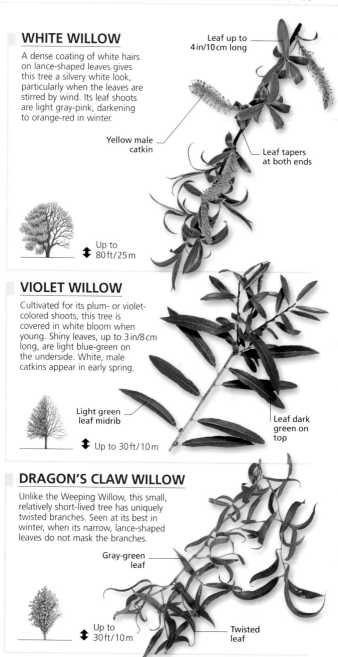

# WHITE WILLOW

A dense coating of white hairs on lance-shaped leaves gives this tree a silvery white look, particularly when the leaves are stirred by wind. Its leaf shoots are light gray-pink, darkening to orange-red in winter.

Leaf up to 4 in/10 cm long

Yellow male catkin

Leaf tapers at both ends

Up to 80 ft/25 m

# VIOLET WILLOW

Cultivated for its plum- or violet-colored shoots, this tree is covered in white bloom when young. Shiny leaves, up to 3 in/8 cm long, are light blue-green on the underside. White, male catkins appear in early spring.

Light green leaf midrib

Leaf dark green on top

Up to 30 ft/10 m

# DRAGON'S CLAW WILLOW

Unlike the Weeping Willow, this small, relatively short-lived tree has uniquely twisted branches. Seen at its best in winter, when its narrow, lance-shaped leaves do not mask the branches.

Gray-green leaf

Twisted leaf

Up to 30 ft/10 m

# Rounded leaves

Trees such as poplars have leaves that are rounded, almost circular, in shape. Poplars may have toothed or slightly lobed leaf margins.

**What to look for** • Tree form • Leaf shape, margin, and arrangement • Leaf stalk • Fall leaf color

## ASPEN

This medium-sized tree is conical when young and spreads broadly with age. Leaves, up to 3 in/7.5 cm long, are borne on slender leaf stalks and "tremble" in the slightest breeze.

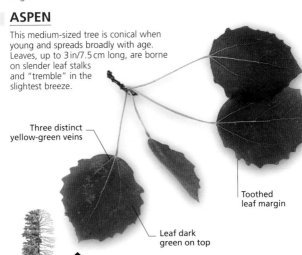

Three distinct yellow-green veins

Toothed leaf margin

Leaf dark green on top

✦ Up to 70 ft/20 m

## GRAY POPLAR

Medium-sized to large hybrid, sometimes forming dense thickets of suckers around the base. Rounded, or sometimes triangular, leaves are held on flattened leaf stalks. Bark has diamond-shaped pits.

Leaf gray-green on top

Grayish white hairs on leaf underside

✦ Up to 100 ft/30 m

BARK

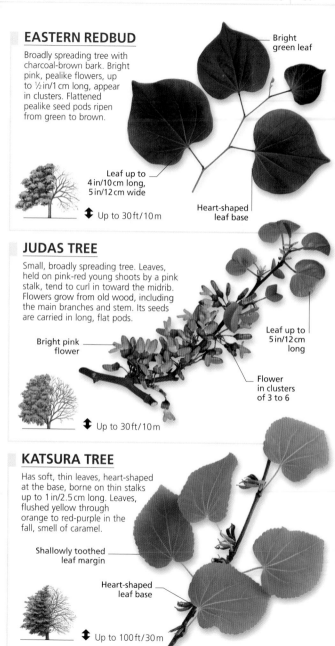

## EASTERN REDBUD

Broadly spreading tree with charcoal-brown bark. Bright pink, pealike flowers, up to ½ in/1 cm long, appear in clusters. Flattened pealike seed pods ripen from green to brown.

Bright green leaf

Leaf up to 4 in/10 cm long, 5 in/12 cm wide

Heart-shaped leaf base

↕ Up to 30 ft/10 m

## JUDAS TREE

Small, broadly spreading tree. Leaves, held on pink-red young shoots by a pink stalk, tend to curl in toward the midrib. Flowers grow from old wood, including the main branches and stem. Its seeds are carried in long, flat pods.

Bright pink flower

Leaf up to 5 in/12 cm long

Flower in clusters of 3 to 6

↕ Up to 30 ft/10 m

## KATSURA TREE

Has soft, thin leaves, heart-shaped at the base, borne on thin stalks up to 1 in/2.5 cm long. Leaves, flushed yellow through orange to red-purple in the fall, smell of caramel.

Shallowly toothed leaf margin

Heart-shaped leaf base

↕ Up to 100 ft/30 m

»

**What to look for** • Tree form • Leaf shape, size, and texture • Leaf color • Flower color • Floral bract

## AMERICAN BASSWOOD

Large tree with large, broadly egg-shaped to rounded leaves, up to 10 in/25 cm long and wide. Leaves have a pointed tip and forward-facing serrations around their margin. Drooping clusters of fragrant flowers appear in early summer.

Leaf deep green on top

Pale yellow flower

Bract along flower cluster

Up to 80 ft/25 m

## SMALL-LEAVED LINDEN

Small, rounded to heart-shaped leaves, up to 3 in/7.5 cm long, are glossy green with a blunt tip. Small, fragrant, pale yellow flowers are borne in pendulous clusters of up to 10, in early summer.

Flower cluster hangs from pale green bract

Blue-green leaf underside

Serrated leaf margin

Up to 100 ft/30 m

## LARGE-LEAVED LINDEN

Large, rounded to broadly oval leaves; deep green with some hairs on top, light green with dense hairs along the midrib and in vein axils on the underside. Fragrant flowers are accompanied by floral bracts, up to 5 in/12.5 cm long.

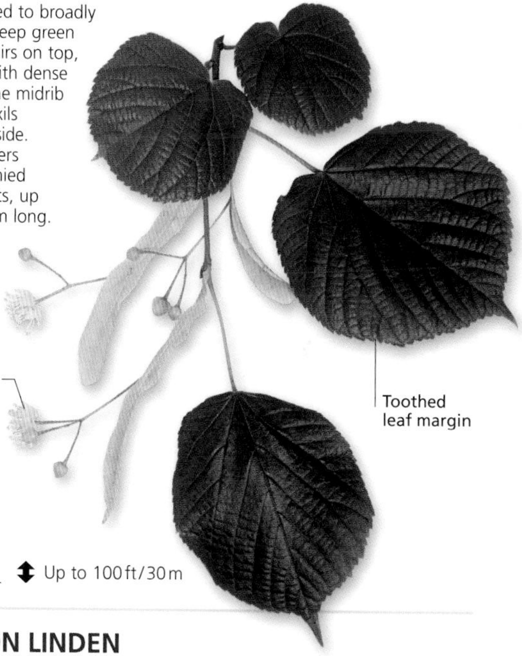
Pale yellow flower, up to ¾ in/2 cm across
Toothed leaf margin

↕ Up to 100 ft/30 m

## COMMON LINDEN

Leaves, heart-shaped at the base, are dull dark green on top, pale green with tufts of hair in the main vein axils on the underside. Old trees have dense suckers around the trunk base.

Pointed leaf tip
Toothed leaf margin
Flower in drooping cluster

↕ 130 ft/40 m

# Tree Life Cycle

Just like humans, trees go through a series of visual changes as they age. A young tree will look very different to an old tree of the same species.

Once planted, a tree may take a few years to start growing. By the time it reaches its "teenage" years, it will grow faster. As the tree ages, its growth rate declines.

## Stages of tree growth

There are several stages of growth during the lifespan of a tree. For the purpose of this book the main stages of tree growth are referred to as "young," "middle-aged," and "old."

**Young tree**
This young Olive still has a relatively compact form, with branches quite low on the stem producing plenty of leaves and vigorous shoots. Trees grow fastest when young.

**Middle-aged tree**
This middle-aged Olive is noticeably taller. Its branches are becoming widespread, and new growth is confined to the top third of the tree.

**Old tree**
As the Olive ages, its lower branches break away from the trunk, and the whole tree begins to lean. In time, the tree may fall over and its branches may die-back.

**CULTIVATED APPLE**

**POCKET HANDKERCHIEF TREE**

## Flowering and fruiting

Trees need to produce flowers before they can bear fruit. However, not all trees produce flowers or fruit when they are young. It may be up to 20 years before the Pocket Handkerchief tree flowers. However, cultivated apple trees may produce fruit from four to five years old.

**WINDSWEPT TREE**

## Climatic conditions

Over time, different climatic conditions can give trees a different shape to their "normal" form. Trees suffering from exposure may become windswept and grow, or lean, away from prevailing winds. Trees growing in areas that receive regular heavy snowfall will become conical in outline, so that snow is shed easily from their branches.

**SNOW-COVERED TREES**

# Elliptic leaves

Several trees, including hollies, have evergreen elliptic leaves, which are widest at or about their middle and taper equally toward both ends.

## KILLARNEY STRAWBERRY TREE

Tough, glossy dark green leaves have a paler underside. Creamy white, urn-shaped flowers appear in drooping clusters. Fruit is inedible and develops from the previous year's flowers, ripening from green to red.

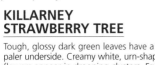

Strawberry-like fruit, up to 1 in/ 2.5 cm across

Serrated leaf margin

↕ Up to 30 ft/10 m

## MADRONE

Broadly columnar tree, with pink-red peeling bark. Has dark green leathery leaves, with a pale blue underside, borne on red stalks. Creamy white, urn-shaped flowers, borne in erect clusters, are followed by strawberry-like fruit.

Leaf up to 6 in/15 cm long, 3 in/7.5 cm wide

↕ Up to 130 ft/40 m

## CHILEAN FIRE BUSH

Small columnar tree, with leaves up to 6 in/15 cm long and 1 in/2.5 cm wide. Branches bear clusters of flowers, each opening into four narrow lobes, which reveal an orange-red style similar to the Honeysuckle.

Matte green leaf

Orange-red tubular flower

↕ Up to 28 ft/9 m

**What to look for** • Tree form • Leaf shape and size
• Leaf color and texture • Flower shape and color
• Fruit shape and color

## COMMON HOLLY

Easily recognizable broadly
columnar, medium-sized
tree. Its rigid, glossy dark green
leaves are waxy on top and
normally display spines around
the margins. Fruit is a round,
bright red berry.

Fruit up to
½ in/1 cm
across

Leaf up to
4 in/10 cm
long

↕ Up to 70 ft/20 m

## AMERICAN HOLLY

Broadly conical tree. Its leaves
are dark matte green on top and
yellow-green on the underside,
with sharp spines around
the margin and at the tip.
Berries appear in winter.

Leaf up to
4 in/10 cm long,
2 in/5 cm wide

Glossy, bright
red berry

↕ Up to 70 ft/20 m

## BAY LAUREL

Conical, densely leaved
tree, with fruity smelling,
leathery leaves that are
used in cooking. Its
yellow-green flowers are
borne in small clusters.
Fruit ripen from
green to black.

Leaf dark
green on top

Oval
berry

↕ Up to 60 ft/18 m

»

**What to look for** • Tree form • Leaf shape, size, and color • Flower shape, size, and color • Fruit size and color

## OLIVE

Small to medium-sized tree, widely cultivated in southern Europe, with broadly spreading branches and a short trunk. Elliptic to inversely egg-shaped leaves are tough and leathery. Edible fruit ripen from green to purple-black.

Fruit up to 1¼in/3cm long

Leaf gray-green on top

Leaf up to 4in/10cm long

Up to 30ft/10m

## HOLM OAK

Large oak, with elliptic to egg-shaped leaves. Leathery, and occasionally toothed, leaves are sage-green with some hairs on the underside. Fruit is a small, pointed acorn held in a fawn-colored, scaly cup.

Leaf dark green on top

Acorn up to ½in/1.5cm long

Up to 100ft/30m

## CHILEAN MYRTLE

Broadly columnar tree with small leaves. Its bright, cinnamon-colored, feltlike bark peels to reveal cream-colored patches. In late summer, it bears fragrant, cup-shaped flowers.

Small white flower

Leaf up to 1 in/2.5 cm long

Red-brown flower stalk

Up to 40 ft/12 m

**BARK**

## BULL BAY

Conical magnolia, with thick leaves, up to 10 in/25 cm long and 4 in/10 cm wide. In summer, large, highly fragrant, ivory-colored flowers are borne singly.

Saucer-shaped flower up to 12 in/30 cm across

Glossy dark green leaf

Up to 70 ft/20 m

# Egg-shaped to elliptic leaves

Trees such as magnolias and some tupelos all bear similar-shaped deciduous leaves, which are either egg-shaped or elliptic.

## PINK TULIP TREE

Broadly conical magnolia, with smooth, gray bark. Leaves, up to 10 in/25 cm long, end in a sharp point. In spring, large, gray, hairy flower buds open to deep pink, waterlily-like flowers up to 12 in/30 cm across.

Spreading outer tepal

Gray-green leaf

Up to 100 ft/30 m

## SPRENGER'S MAGNOLIA

Broadly conical tree. Similar to the Pink Tulip Tree but bears smaller flowers. Rose-pink and goblet-shaped flowers fade to light pink, and open to 8 in/20 cm across.

Unripe green fruit

Limp tepal

Up to 90 ft/28 m

## SAUCER MAGNOLIA

Common garden hybrid, with a short trunk and widely spreading branches. Produces large, goblet-shaped flowers, up to 8 in/20 cm tall, in early spring before the leaves appear.

Dark green leaf

Pink-purple staining at tepal base

Up to 40 ft/12 m

**What to look for** • Tree form • Leaf shape and size • Fall leaf color • Flower shape, size, and color

## STAR MAGNOLIA

Slow-growing and broadly spreading magnolia. Its leaves are up to 4 in/10 cm long and 2 in/5 cm wide, with a pale green underside. Flowers, with up to 20 narrow white or pink-blushed petals, open in spring.

Deep green leaf

Starlike flower

Up to 12 ft/4 m

## WILSON'S MAGNOLIA

Broadly spreading, multi-stemmed, shrublike tree. Leaves run to a pointed tip. In midspring, silver-haired buds release fragrant flowers with crimson-purple stamens.

Leaf up to 8 in/ 20 cm long

Drooping, cup-shaped white flower

Up to 25 ft/8 m

## BLACK TUPELO

Broadly conical tree native to North America. Its slender branches bear bright green leaves, up to 6 in/15 cm long. Grown mainly for its fall leaf tints, which range from vivid yellow-orange to burgundy-red.

Tiny green flower

Fall leaf

Up to 80 ft/25 m

Blunt leaf tip

»

**What to look for** • Tree form • Leaf shape and size • Flower shape, size, and arrangement • Fruit shape, size, and color

## WILD CHERRY

A common tree, also known as the Gean. White, five-petaled, fragrant flowers appear in drooping clusters in spring. Bark is red-gray with a slight sheen, and is marked by shallow, horizontal cracks.

Leaf up to 6 in/15 cm long

Flower up to 1¼ in/3 cm across

↕ Up to 80 ft/25 m

## PLUM

Broadly spreading tree widely grown in gardens for its fleshy, purple-skinned, sweet-tasting fruit. Fragrant white flowers are borne on bare branches in spring, before elliptic, dull green leaves appear.

Leaf up to 3 in/7.5 cm long

Fruit up to 2 in/5 cm across

↕ Up to 30 ft/10 m

## BIRD CHERRY

Common small to medium-sized tree, recognized by its long clusters of fragrant flowers. Bears distinct round to oval-shaped berries, which ripen from red to black and are a preferred food source for birds.

Leaf up to 4 in/10 cm long

Small white flower

↕ Up to 50 ft/15 m

## SARGENT'S CHERRY

Attractive, broadly spreading
cherry, which produces
masses of rose-pink
flowers in spring, just as
its bronze-colored leaves
appear. Leaves turn deep
orange in the fall. Surface
roots often spread out
from the base of
the trunk.

Flower
1¼–1½ in/3–4 cm
across

Notched
flower petal

🔼 Up to 70 ft/20 m

## KANZAN
## FLOWERING CHERRY

Widely planted ornamental
cherry. Grown for its large
clusters of semidouble flowers,
borne in midspring. At other
times it is recognized by its
spreading habit and dark green
leaves with a serrated margin.

Bright pink
flower

🔼 Up to 40 ft/12 m

## GREAT WHITE CHERRY

Beautiful Japanese cherry of
garden origin. Large, dazzling
white, single flowers appear
in midspring, as do young,
bronze-colored leaves. These
fade to green and then to
orange-red in the fall.

Flower up to
3 in/8 cm across

🔼 Up to 30 ft/10 m

»

**What to look for** • Tree form • Leaf shape, size, and color • Flower color • Fruit size and color

## MOUNT FUJI CHERRY

Small cherry tree, with long, arching, almost horizontal branches. In spring, it is covered in large, semidouble flowers hanging in long-stemmed clusters, below bright green, unfurling leaves.

Serrated leaf margin

Snow-white flower

↕ Up to 20 ft/6 m

## BLACKTHORN

Often found growing as a hedgerow, with a dense tangle of heavily spined branches. Small white flowers appear in midspring. Berrylike fruit, called sloes, are produced from late summer to fall.

Blue-black fruit

Leaf up to 1½ in/4 cm long

↕ Up to 30 ft/10 m

Toothed leaf margin

## COMMON PEAR

Believed to be an ancient hybrid from which many domestic fruiting pears have been cultivated. Produces clusters of white, five-petaled flowers with purple-red anthers in spring. Pears appear in late summer.

Leaf glossy dark green on top

Fruit up to 4 in/10 cm long

↕ Up to 50 ft/15 m

## WHITEBEAM

Broadly columnar tree. Its leaves have irregular, shallow serrations around the margins, and are covered in dense, silver-white hairs on the undersides. Produces glossy red berries in late summer.

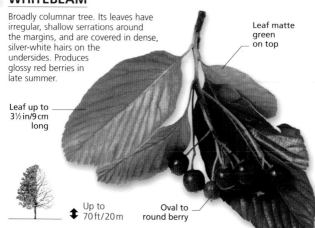

Leaf matte green on top

Leaf up to 3½ in/9 cm long

Up to 70 ft/20 m

Oval to round berry

## SWEDISH WHITEBEAM

Small, extremely hardy tree. Identified by its leaves, which are similar to the Whitebeam. However, its leaf margins have indentations, sometimes deep toward the leaf midrib.

Irregularly lobed leaf margin

Pale green leaf underside

Up to 50 ft/15 m

Deep red, oval berry

## DUTCH ELM

Large, broadly columnar hybrid, found throughout Western Europe. Its egg-shaped leaves are rough to the touch, unequal at the base, and covered with hairs on the underside.

Green-gray leaf

Serrated leaf margin

Up to 100 ft/30 m

»

**What to look for** • Tree form • Leaf shape and color
• Flower size and color • Fruit color • Type of branching

## VARIEGATED TABLE DOGWOOD

Widely grown cultivar of the Table Dogwood,
this tree has the same symmetrical, horizontal
branching, but is smaller and grows more slowly.
Narrow, variegated leaves have striking margins.

Narrow,
variegated
leaf

Broad,
silver-cream
leaf margin

↕ Up to 25 ft/8 m

## PACIFIC DOGWOOD

Medium-sized, multiple-stemmed tree. In
late spring, it bears clusters of tiny flowers,
surrounded by four to seven petal-like bracts,
each up to 3 in/7.5 cm long. These are
followed by pink, strawberry-like fruit.

Creamy white,
often blushed
pink, bract

Leaf dark
green on top

Up to
↕ 80 ft/25 m

Untoothed
leaf margin

## WEDDING-CAKE TREE

Broadly spreading tree, recognized by its symmetrical, horizontal branches, which gradually grow shorter toward the top, like the layers of a wedding cake. Flattened flowerheads, borne in early summer, are followed by clusters of small, blue-black fruit carried on pink-red stalks.

Leaf shiny, dark green on top

Creamy white flower

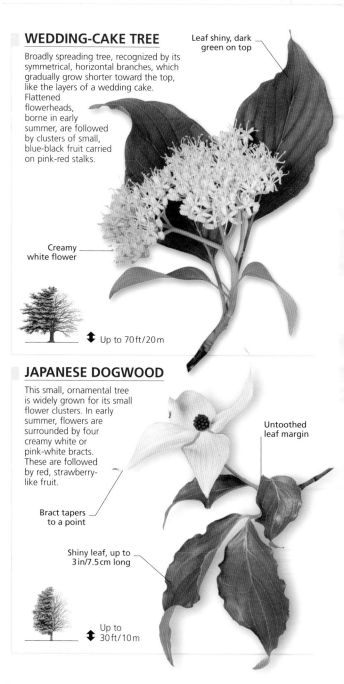

Up to 70 ft/20 m

## JAPANESE DOGWOOD

This small, ornamental tree is widely grown for its small flower clusters. In early summer, flowers are surrounded by four creamy white or pink-white bracts. These are followed by red, strawberry-like fruit.

Untoothed leaf margin

Bract tapers to a point

Shiny leaf, up to 3 in/7.5 cm long

Up to 30 ft/10 m

# Elliptic to inversely egg-shaped leaves

These elm trees have leaves that are widest at, or just above, their middle. In late summer, they produce paperlike disks that contain a single seed.

**What to look for** • Tree form • Leaf size, shape, and color
• Leaf base and margin • Fruit color

## EUROPEAN WHITE ELM

Widely spreading tree, with inversely egg-shaped leaves up to 4½in/11 cm long and 2½in/6 cm wide. Oval, paperlike fruit ripens from green to brown and contains a single seed. Its bark remains smooth even in old age.

Leaf mid-green on top

Double-toothed leaf margin

Asymmetrical leaf base

Up to 100 ft/30 m

## WYCH ELM

Large tree, with a natural range extending from Portugal to Russia. Leaves, up to 8 in/20 cm long and 4 in/10 cm wide, are dull green and rough on top, paler with gray hairs on the underside. Distinctly veined, they are asymmetrical at the base.

Double-toothed leaf margin

Bright yellow leaf in the fall

Up to 100 ft/30 m

# Triangular leaves

Trees including some poplars and the Silver Birch have deciduous triangular leaves, which end in an acute point at the tip.

**What to look for** • Tree form • Leaf size, shape, and color • Leaf stalk length and color • Bark color

## BLACK POPLAR

Large, fast-growing tree with pale gray bark. Its leaves are coated with silver-gray hairs on the underside and are borne on a long, flattened leaf stalk.

Glossy green leaf

Leaf up to 3 in/8 cm long and wide

↕ Up to 100 ft/30 m

## LOMBARDY POPLAR

Distinctive tall, slender, columnlike outline and strongly upcurved branches. A subspecies of the Black Poplar, its leaves are glossy green on top and pale green on the underside.

Leaf up to 2½ in/6 cm long and wide

Yellow-green stalk

↕ Up to 100 ft/30 m

## SILVER BIRCH

Leaf tapers to point

Slender-trunked tree, with white or silver-gray bark and drooping, light, and airy branch structure. Leaves, up to 2½ in/6 cm long and 1½ in/4 cm wide, are irregularly toothed around the margin.

Fruiting catkin, up to 1¼ in/3 cm long

↕ Up to 100 ft/30 m

BARK

# Oblong leaves

Trees such as the Caucasian Elm and the Chinese Tupelo have oblong leaves that are longer than they are broad, with sides that are parallel or almost so.

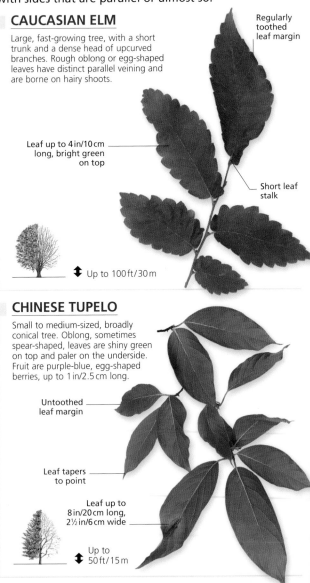

## CAUCASIAN ELM

Large, fast-growing tree, with a short trunk and a dense head of upcurved branches. Rough oblong or egg-shaped leaves have distinct parallel veining and are borne on hairy shoots.

Regularly toothed leaf margin

Leaf up to 4 in/10 cm long, bright green on top

Short leaf stalk

Up to 100 ft/30 m

## CHINESE TUPELO

Small to medium-sized, broadly conical tree. Oblong, sometimes spear-shaped, leaves are shiny green on top and paler on the underside. Fruit are purple-blue, egg-shaped berries, up to 1 in/2.5 cm long.

Untoothed leaf margin

Leaf tapers to point

Leaf up to 8 in/20 cm long, 2½ in/6 cm wide

Up to 50 ft/15 m

**What to look for** • Tree form • Leaf shape, size, and color • Leaf margin, tip, and veining • Fruit size and color

## RAULI SOUTHERN BEECH

Fast-growing, broadly conical tree native to South America. Its deep green leaves taper to a blunt tip and are clearly marked by 14 to 18 pairs of parallel veins, extending from a yellow-green midrib.

Ripe husk, up to ½in/1cm long

Leaf up to 4in/10cm long, 2in/5cm wide

Up to 80ft/25m

Toothed leaf margin

## SWEET CHESTNUT

Long-lived, broadly columnar tree. Its leaves, up to 8in/20cm long, are deep green on top and paler on the underside. Fruit is an edible chestnut encased in a spiky husk.

Coarsely toothed leaf margin

Yellow-green husk

Up to 100ft/30m

Creamy yellow flower on erect catkin

»

**What to look for** • Tree form • Leaf shape, size, and color
• Leaf margin, tip, and veining • Flower size and color

## ULMO

Broadly columnar tree native
to South America. Stiff leaves
are wavy-edged and have
blunt serrations around the
margin. Fragrant flowers,
with purple-tipped orange
stamens, are profusely
borne in late summer.

Flower up to
1½ in/4 cm across

Leaf dark
green on top

Ivory-colored
flower

Up to
70 ft/20 m

## KOHUHU

Columnar tree, with oblong,
sometimes elliptic, wavy-
edged leaves up to 2 in/5 cm
long and 1 in/2.5 cm wide.
Leaves have a white
midrib and are pale green
to white on
the underside.

Reddish purple
flower

Waxy leaf
shiny bright
green on top

Up to 25 ft/8 m

# Palmlike leaves

Some treelike plants such as the Chusan Palm have palmlike leaves that are divided almost to the leaf stalk. Each leaf can consist of up to 40 long, thin strips.

**What to look for** • Leaf shape, size, and arrangement
• Flower color and arrangement • Bark color and texture

## CHUSAN PALM

Palm with stiff, leathery leaves, up to 4 ft/1.2 m wide, attached to the trunk by stiff, flattened leaf stalks. Gray-brown, fibrous-haired bark is marked by the scars of discarded leaves.

Creamy yellow male flower

Leaf stalk up to 3 ft/1 m long

Dark green leaf

Up to 40 ft/12 m

# 3 BROADLEAVES: COMPOUND

Trees with leaves that are divided into leaflets are said to have compound leaves. The leaflets may be arranged along an axis, as in a feather, or may radiate from the same point. On some leaves, the leaflets themselves are subdivided.

**HONEY LOCUST**
*p.97*

## Leaf shapes

Leaves with subdivided leaflets can look fernlike. Those with four or five leaflets can look like a hand. Some trees, such as laburnums, have leaves with just three leaflets.

**BLACK LOCUST**
*p.94*

**COMMON HORSE CHESTNUT**
*p.100*

**COMMON LABURNUM**
*p.102*

# Leaves with leaflets

The leaves of these trees are subdivided into smaller leaflets. These leaflets are arranged on either side of a stalk, called a rachis.

## TREE OF HEAVEN

Broadly columnar tree. Leaves, up to 30 in/75 cm long, consist of 11 to 19 leaflets. Each leaflet, up to 5½ in/15 cm long, has a small number of serrations at its base. The yellow-brown winged seed is tinged with red when ripe.

Elongated tip of leaflet

Leaflet dark green on top

↕ Up to 100 ft/30 m

## COMMON ASH

Recognized by its distinctive, black leaf buds, evident through winter; its clusters of purple-yellow flowers in early spring; and its winged seeds from summer until winter.

Shallow-toothed leaf margin

↕ Up to 130 ft/40 m

Drooping seed bunch

## FLOWERING ASH

Broadly spreading tree, with brown buds—shaped like a bishop's miter—from fall until spring. Fluffy, open clusters of fragrant flowers are borne in spring. Leaves have five to nine leaflets, each up to 4 in/10 cm long.

Creamy white flowers

↕ Up to 70 ft/20 m

Leaf matte green on top

**What to look for** • Tree form • Number of leaflets in leaf • Flower arrangement and color • Fruit color

## PRIDE OF INDIA

Medium-sized, broadly spreading tree. Large leaves, up to 18 in/45 cm long, have 11 to 13 leaflets with toothed margins, which give the foliage a fernlike appearance. Erect clusters of golden yellow flowers appear in summer.

Flower up to ½ in/1 cm long

Deeply lobed leaflet

Up to 40 ft/12 m

## BLACK WALNUT

Large, fast-growing tree, native to North America. Large leaves have up to 17 leaflets, with toothed margins and pointed tips. Leaves are dark glossy green on top, covered in hairs on the underside.

Leaflet up to 4 in/10 cm long

Male catkin

Up to 100 ft/30 m

## ENGLISH WALNUT

Slow-growing, medium-sized, broadly spreading tree, with light gray bark. Leaves have up to nine, broadly egg-shaped leaflets, each up to 5 in/13 cm long. Fruit is a round green husk, encasing the familiar edible walnut.

Leaflet up to 5 in/13 cm long

Fruit up to 2 in/5 cm across

Up to 100 ft/30 m

**What to look for** • Tree form • Leaf shape and arrangement
• Number of leaflets in leaf • Flower color • Fruit color

## BLACK LOCUST

Common in towns and
cities; its distinctive gray-
brown bark is cracked
and vertically ridged.
Leaves, up to 12 in/
30 cm long, have 11
to 21 leaflets. Pealike,
fragrant white flowers
appear in summer.

Limp leaflet,
up to 2 in/5 cm
long

Flower up to
¾ in/2 cm long

Up to 80 ft/25 m

## GOLDEN-LEAVED ROBINIA

This smaller and less vigorous version
of the Black Locust has attractive,
brightly colored, golden yellow
leaves, borne from spring to early
fall. Very popular for planting in
town parks and gardens.

Oval-shaped
leaflet

Thin, soft
leaflet

Up to 50 ft/15 m

## COMMON ELDER

Multi-stemmed small tree or
large shrub. Leaves, up to
12 in/30 cm long, have five
to nine leaflets. Pungent-
smelling flowers in summer
are followed by drooping
clusters of small, purple-
black berries.

Creamy
white flower

Leaflet 2–5 in/
5–12 cm long

Up to 20 ft/6 m

# ROWAN

Broadly conical tree also known as the Mountain Ash. Leaves, up to 8 in/20 cm long, carry up to 15 leaflets. Clusters of creamy white flowers in spring are followed by bunches of bright red berries.

Sharp-toothed leaf margin

Berry up to ½ in/1 cm across

Up to 70 ft/20 m

# AMERICAN MOUNTAIN ASH

Broadly spreading tree, with leaves up to 10 in/25 cm long. Each leaf has up to 15 lance-shaped, pointed leaflets. Flowers are followed by bright red berries.

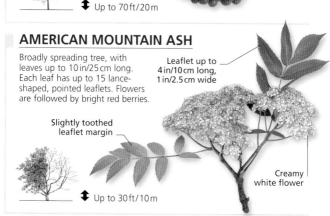

Leaflet up to 4 in/10 cm long, 1 in/2.5 cm wide

Slightly toothed leaflet margin

Creamy white flower

Up to 30 ft/10 m

# JAPANESE ROWAN

Distinctive tree identified by its long-pointed, red winter buds. Leaves up to 12 in/30 cm long, have 13 to 15 lance-shaped leaflets. Flowers are followed by orange-red fruit in the fall.

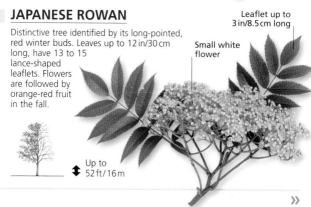

Leaflet up to 3 in/8.5 cm long

Small white flower

Up to 52 ft/16 m

»

**What to look for** • Tree form • Number of leaflets in leaf
• Bark texture

## BITTERNUT HICKORY

Broadly columnar, with leaves
comprising up to four pairs of
opposite leaflets and one terminal
leaflet. Green, oval fruit consists
of a thin shell encasing a
bitter-tasting nut.

Male catkin up to
3 in/7.5 cm long

Leaflet up to
6 in/15 cm long

↕ Up to 100 ft/30 m

## SHAGBARK HICKORY

Large, broadly columnar tree.
Its shaggy gray bark curls away
from the tree in thin strips, up
to 12 in/30 cm long, remaining
attached to the tree at their center.
Leaves have five to seven leaflets,
each with a long tip.

Leaflet up to
4 in/10 cm long

Male
catkin

↕ Up to 100 ft/30 m

## BIG-BUD HICKORY

Leaflet up to
12 in/30 cm
long

Also known as the
Mockernut, its large leaf
bud is up to ¾ in/2 cm long.
Leaves with up to nine leaflets
are dark green on top with
dense, short hairs on the underside.
Silver-gray bark develops vertical
orange fissures with age.

Egg-shaped
leaflet with
upcurved tip

↕ Up to 100 ft/30 m

# Leaves with divided leaflets

The compound leaves on these trees are divided into leaflets, which are further divided. This makes the foliage appear feathery.

**What to look for** • Tree form • Leaf shape and arrangement • Leaf color • Number of leaflets in leaf

## SILVER WATTLE

Medium-sized tree, also known as the Mimosa. Prized by florists for its delicate, feathery leaves and its fragrant, small, pom-pom-like flowers, borne on clusters, up to 4 in/10 cm long.

Sulfur-yellow flower, up to ½ in/1 cm across

Sage-green leaflet

↕ Up to 80 ft/25 m

## HONEY LOCUST

Broadly spreading tree, identified by ferocious thorns covering its trunk and main branches. Leaves consist of leaflets, which are sometimes further divided. Each leaflet has up to 18 pairs of smaller leaflets.

Leaflet glossy green on top

Leaflet up to 1½ in/4 cm long

↕ Up to 100 ft/30m

BARK

# Winter Twigs and Buds

It may be difficult to identify deciduous broadleaf trees in winter, with no leaves to view. However, their twigs and buds can offer several clues.

Winter buds form on shoots, and are normally positioned at the extremities of the previous year's growth. They become visible once the leaves have dropped in the fall. The buds contain all that is needed to commence either leaf or flower growth during the following spring.

### Common Ash
The winter buds of the Common Ash are easy to recognize because they are black and velvety to the touch. They stand out clearly in contrast to the light gray bark of the surrounding twigs.

Black leaf bud

Lime-green terminal bud

### Sycamore
This well-known tree can be identified in winter by its lime-green buds. These buds are borne in opposite pairs along the twigs, with one larger terminal bud at the tip.

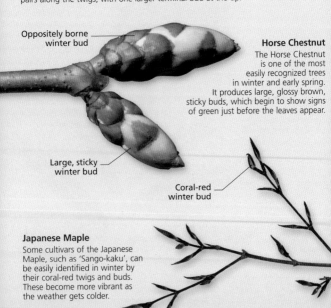

Oppositely borne winter bud

### Horse Chestnut
The Horse Chestnut is one of the most easily recognized trees in winter and early spring. It produces large, glossy brown, sticky buds, which begin to show signs of green just before the leaves appear.

Large, sticky winter bud

Coral-red winter bud

### Japanese Maple
Some cultivars of the Japanese Maple, such as 'Sango-kaku', can be easily identified in winter by their coral-red twigs and buds. These become more vibrant as the weather gets colder.

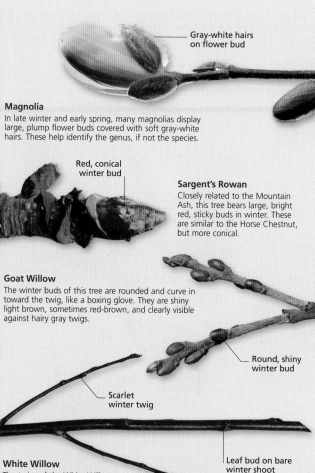

Gray-white hairs
on flower bud

### Magnolia
In late winter and early spring, many magnolias display large, plump flower buds covered with soft gray-white hairs. These help identify the genus, if not the species.

Red, conical
winter bud

### Sargent's Rowan
Closely related to the Mountain Ash, this tree bears large, bright red, sticky buds in winter. These are similar to the Horse Chestnut, but more conical.

### Goat Willow
The winter buds of this tree are rounded and curve in toward the twig, like a boxing glove. They are shiny light brown, sometimes red-brown, and clearly visible against hairy gray twigs.

Round, shiny
winter bud

Scarlet
winter twig

### White Willow
The twigs of the White Willow are green-yellow for most of the year, but turn orange-yellow in late winter. In some cultivars, such as 'Britzensis' they can be vibrant scarlet.

Leaf bud on bare
winter shoot

Orange-brown
leaf bud

### English Oak
The leaf buds of the English Oak have a blunt or rounded tip. Orange-brown and up to ¼ in (5 mm) long, they are typically clustered around the twig tips.

# Hand-shaped, divided leaves

These trees have leaves with indentations that cut down to the main leaf stalk, creating separate leaflets. They produce a husk, which contains seeds known as conkers.

## CALIFORNIA BUCKEYE

Broadly spreading, multi-stemmed tree. Leaves have five to seven leaflets. Its pale pink flowers are borne in upright, columnar clusters. Seed is a pear-shaped, glossy brown conker protected by a green casing.

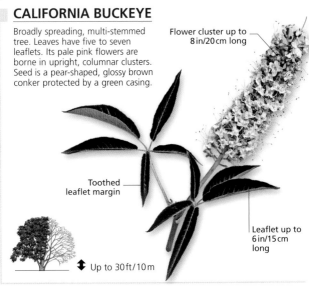

Flower cluster up to 8 in/20 cm long

Toothed leaflet margin

Leaflet up to 6 in/15 cm long

↕ Up to 30 ft/10 m

## COMMON HORSE CHESTNUT

One of the most common large trees in cultivation. Its leaves consist of up to seven egg-shaped leaflets with a narrow base, each up to 10 in/25 cm long. Creamy white flowers speckled with pink appear in spring.

Strong veins

Bright green leaflet

↕ Up to 100 ft/30 m

**What to look for** • Tree form • Leaf color • Number of leaflets in leaf • Leaflet shape and size • Flower color

## INDIAN HORSE CHESTNUT

Broadly columnar tree, which flowers later than the Common Horse Chestnut. Upright clusters of white to pale pink flowers, up to 10 in/25 cm long, appear in early summer. Its leaves have five to seven glossy green leaflets, each up to 10 in/25 cm long.

Rough green fruit

Stout stalk bears fruit

Lance-shaped leaflet

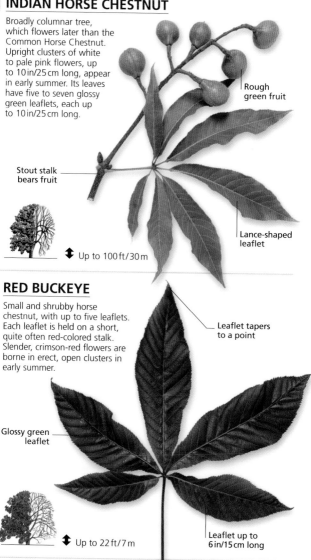

↕ Up to 100 ft/30 m

## RED BUCKEYE

Small and shrubby horse chestnut, with up to five leaflets. Each leaflet is held on a short, quite often red-colored stalk. Slender, crimson-red flowers are borne in erect, open clusters in early summer.

Leaflet tapers to a point

Glossy green leaflet

↕ Up to 22 ft/7 m

Leaflet up to 6 in/15 cm long

# Trifoliate leaves

Some trees such as the laburnums have trifoliate leaves, which consist of three separate leaflets. Clover is another widely found plant with this leaf type.

## COMMON LABURNUM

One of the most familiar small trees in cultivation. Easily recognized by its soft leaves, which have three leaflets, each up to 4 in/10 cm long. Flowers, up to 1 in/2.5 cm long, are borne in drooping clusters. Green, beanlike seed pods contain poisonous seeds.

Pealike flowers in clusters up to 12 in/30 cm long

Leaflet rich green on top

Golden yellow flower

Up to 28 ft/9 m

## VOSS'S LABURNUM

Broadly spreading tree; produces larger clusters of pealike flowers in greater numbers than the Common Laburnum, making it more popular for planting in parks and gardens. It also has slightly hairier leaves and produces fewer poisonous seeds.

Leaflet deep green on top

Golden yellow flower

Flower cluster up to 18 in/45 cm long

 Up to 50 ft/15 m

**What to look for** • Tree form • Number of leaflets in leaf
• Leaflet shape, size, and color • Flower size and color
• Fruit shape • Bark color and texture

## PAPERBARK MAPLE

Ornamental tree identified
by its cinnamon-red bark,
which is smooth when
young but peels in papery
flakes with age. Leaves
have three elliptic leaflets,
with blunt-toothed margins.

Leaflet dark
green on top

Blue-green
leaf underside

Pale green,
winged fruit

↕ Up to 50 ft/15 m

**BARK**

## NYMAN'S EUCRYPHIA

Small, upright hybrid between
the Eucryphia and the Ulmo
(p.88). In late summer,
masses of fragrant flowers,
up to 3 in/7.5 cm across,
appear. Leaves, up to 2½ in/
6 cm long, may have three
leaflets, and
are paler on
the underside.

Toothed leaf
margin

Leaflet glossy
dark green
on top

↕ Up to 50 ft/15 m

Four-petaled,
ivory-white
flower

# TREE GALLERY

This gallery shows the trees profiled in the book, grouped by family. A tree family consists of closely related genera, which are, in turn, made up of related species that have common reproductive features—flowers or cones. Use this section to identify related trees and go to their profile pages to learn how they differ from each other.

**ACERACEAE**
*Maples*

Field Maple
*p.42*

Paperbark Maple
*p.103*

Japanese Maple
*p.42*

Norway Maple
*p.42*

Sugar Maple
*p.43*

Red Maple
*p.43*

Sycamore
*p.43*

**ADOXACEAE**
*Elders*

Common Elder
*p.94*

**AQUIFOLIACEAE**
*Hollies*

Common Holly
*p.73*

American Holly
*p.73*

**ARAUCARIACEAE**
*Monkey Puzzle
and Wollemi Pine*

Monkey Puzzle
*p.38*

Wollemi Pine
*p.23*

**BETULACEAE**
*Alders, birches,
hornbeams, and hazels*

Italian Alder
*p.51*

Common Alder
*p.49*

Gray Alder
*p.51*

Chinese Red-barked Birch
*p.50*

Paper-bark Birch
*p.53*

Silver Birch
*p.85*

Downy Birch
*p.50*

Himalayan Birch
*p.50*

Common Hornbeam
*p.52*

Turkish Hazel
*p.53*

## BIGNONIACEAE
*Indian Bean Tree*

Indian Bean Tree
*p.52*

## CERCIDIPHYLLACEAE
*Katsuras*

Katsura Tree
*p.67*

## CORNACEAE
*Dogwoods*

Wedding-cake Tree
*p.83*

Variegated Table
Dogwood
*p.82*

Japanese Dogwood
*p.83*

Pacific Dogwood
*p.82*

## CUPRESSACEAE
*Cypress trees
and junipers*

Incense Cedar
*p.39*

Western Red Cedar
*p.39*

Leyland Cypress
*p.37*

Monterey Cypress
*p.37*

Italian Cypress
*p.36*

Lawson Cypress
*p.37*

Common Juniper
*p.33*

**DICKSONIACEAE**
*Soft Tree Fern*

Soft Tree Fern
*p.61*

**ERICACEAE**
*Madrone and
strawberry trees*

Madrone
*p.72*

Killarney
Strawberry Tree
*p.72*

**EUCRYPHIACEAE**
*Eucryphias*

Ulmo
*p.88*

Nyman's Eucryphia
*p.103*

**FAGACEAE**
*Chestnuts, beeches,
and oaks*

Sweet Chestnut
*p.87*

Oriental Beech
*p.55*

Common Beech
*p.55*

Purple Beech
*p.55*

Rauli Southern Beech
*p.87*

»

»

Antarctic Beech
*p.58*

Coigüe Southern Beech
*p.60*

Roble Beech
*p.58*

Turkey Oak
*p.47*

Scarlet Oak
*p.47*

Holm Oak
*p.74*

Sessile Oak
*p.46*

Pedunculate Oak
*p.47*

Red Oak
*p.46*

Cork Oak
*p.60*

**GINKGOACEAE**
*Maidenhair Tree*

Maidenhair Tree
*p.35*

**HAMAMELIDACEAE**
*Sweetgum and
Persian Ironwood*

Sweetgum
*p.44*

Persian Ironwood
*p.49*

**HIPPOCASTANACEAE**
*Horse chestnuts*

Common Horse Chestnut
*p.100*

Indian Horse Chestnut
*p.101*

Red Buckeye
*p.101*

California Buckeye
*p.100*

**JUGLANDACEAE**
*Hickories and walnuts*

Bitternut Hickory
*p.96*

Shagbark Hickory
*p.96*

Big-bud Hickory
*p.96*

Black Walnut
*p.93*

English Walnut
*p.93*

**LAURACEAE**
*Bay Laurel*

Bay Laurel
*p.73*

## LEGUMINOSAE
*Laburnums and locust trees*

**Eastern Redbud**
*p.67*

**Judas Tree**
*p.67*

**Honey Locust**
*p.97*

**Black Locust**
*p.94*

**Golden-leaved Robinia**
*p.94*

**Common Laburnum**
*p.102*

**Voss's Laburnum**
*p.102*

**Silver Wattle**
*p.97*

## MAGNOLIACEAE
*Magnolias and tulip trees*

**Tulip Tree**
*p.48*

**Pink Tulip Tree**
*p.76*

**Bull Bay**
*p.75*

**Sprenger's Magnolia**
*p.76*

**Star Magnolia**
*p.77*

Wilson's Magnolia
*p.77*

Saucer Magnolia
*p.76*

**MORACEAE**
*Mulberries*

White Mulberry
*p.56*

Black Mulberry
*p.56*

**MYRTACEAE**
*Gums and myrtles*

Tasmanian Blue Gum
*p.63*

Cider Gum
*p.63*

Chilean Myrtle
*p.75*

**NYSSACEAE**
*Tupelos and Pocket Handkerchief Tree*

Pocket
Handkerchief Tree
*p.53*

Chinese Tupelo
*p.86*

Black Tupelo
*p.77*

### OLEACEAE
*Olive and ash trees*

Common Ash
*p.92*

Flowering Ash
*p.92*

Olive
*p.74*

### PALMAE
*Chusan Palm*

Chusan Palm
*p.89*

### PINACEAE
*Firs, cedars, larches, spruces, and pines*

Grand Fir
*p.20*

Caucasian Fir
*p.21*

Noble Fir
*p.21*

Douglas Fir
*p.23*

Deodar Cedar
*p.33*

Cedar of Lebanon
*p.32*

Atlas Cedar
*p.32*

Japanese Larch
*p.34*

**Hybrid Larch**
*p.34*

**European Larch**
*p.34*

**Norway Spruce**
*p.27*

**Brewer's Weeping Spruce**
*p.26*

**Serbian Spruce**
*p.27*

**Oriental Spruce**
*p.27*

**Sitka Spruce**
*p.26*

**Western Yellow Pine**
*p.31*

**Austrian Pine**
*p.30*

**Monterey Pine**
*p.30*

**Bhutan Pine**
*p.31*

**Scots Pine**
*p.31*

**Western Hemlock**
*p.23*

**PITTOSPORACEAE**
*Kohuhu*

**Kohuhu**
*p.88*

**PLATANACEAE**
*Plane trees*

Buttonwood
*p.45*

Oriental Plane
*p.45*

London Plane
*p.45*

**PROTEACEAE**
*Chilean Fire Bush*

Chilean Fire Bush
*p.72*

**ROSACEAE**
*Hawthorns, crab apples, cherries, plums, pears, apples, and rowans*

Snowy Mespil
*p.51*

Midland Hawthorn
*p.48*

Hawthorn
*p.48*

Siberian Crab Apple
*p.57*

Common Crab Apple
*p.57*

Cultivated Apple
*p.57*

Common Pear
*p.80*

Willow-leaved Pear
*p.64*

Portuguese Laurel
*p.60*

Peach
*p.64*

Plum
*p.78*

Apricot
*p.59*

Bird Cherry
*p.78*

Wild Cherry
*p.78*

Sargent's Cherry
*p.79*

Tibetan Cherry
*p.64*

Kanzan Flowering Cherry
*p.79*

Mount Fuji Cherry
*p.80*

Great White Cherry
*p.79*

Blackthorn
*p.80*

Rowan
*p.95*

Japanese Rowan
*p.95*

American
Mountain Ash
*p.95*

»

»

Whitebeam
*p.81*

Swedish Whitebeam
*p.81*

**SALICACEAE**
*Poplars and willows*

White Poplar
*p.44*

Chinese Necklace
Poplar
*p.59*

Black Poplar
*p.85*

Lombardy Poplar
*p.85*

Gray Poplar
*p.66*

Aspen
*p.66*

White Willow
*p.65*

Dragon's Claw Willow
*p.65*

Violet Willow
*p.65*

Crack Willow
*p.62*

Golden
Weeping Willow
*p.62*

Common Osier
*p.62*

**SAPINDACEAE**
*Pride of India*

Pride of India
*p.93*

**SCROPHULARIACEAE**
*Empress Tree*

Empress Tree
*p.59*

**SIMAROUBACEAE**
*Tree of Heaven*

Tree of Heaven
*p.92*

**STYRACACEAE**
*Mountain Snowdrop Tree*

Mountain
Snowdrop Tree
*p.56*

**TAXACEAE**
*Yews*

Common Yew
*p.22*

**TAXODIACEAE**
*Cedars and redwoods*

Japanese Cedar
*p.39*

Dawn Redwood
*p.24*

Coastal Redwood
*p.22*

Giant Redwood
*p.38*

Swamp Cypress
*p.25*

**TILIACEAE**
*Lindens*

American Basswood
*p.68*

Small-leaved Linden
*p.68*

»

》

Large-leaved Linden
*p.69*

Common Linden
*p.69*

**ULMACEAE**
*Elms, Hackberry,
and Keyaki*

Hackberry
*p.52*

Wych Elm
*p.84*

European White Elm
*p.84*

English Elm
*p.54*

Dutch Elm
*p.81*

Caucasian Elm
*p.86*

Keyaki
*p.54*

# Scientific Names

All trees have a unique scientific name of two Latin words that represent its genus and specific name. Cultivars and hybrids carry additional names to set them apart from the parent. Hybrids have an "x" in their name, indicating a cross between two species.

| Common name | Scientific name | Page |
|---|---|---|
| Grand Fir | *Abies grandis* | 20 |
| Caucasian Fir | *Abies nordmanniana* | 21 |
| Noble Fir | *Abies procera* | 21 |
| Common Yew | *Taxus baccata* | 22 |
| Coastal Redwood | *Sequoia sempervirens* | 22 |
| Douglas Fir | *Pseudotsuga menziesii* | 23 |
| Western Hemlock | *Tsuga heterophylla* | 23 |
| Wollemi Pine | *Wollemia nobilis* | 23 |
| Dawn Redwood | *Metasequoia glyptostroboides* | 24 |
| Swamp Cypress | *Taxodium distichum* | 25 |
| Sitka Spruce | *Picea sitchensis* | 26 |
| Brewer's Weeping Spruce | *Picea breweriana* | 26 |
| Serbian Spruce | *Picea omorika* | 27 |
| Oriental Spruce | *Picea orientalis* | 27 |
| Norway Spruce | *Picea abies* | 27 |
| Austrian Pine | *Pinus nigra* | 30 |
| Monterey Pine | *Pinus radiata* | 30 |
| Scots Pine | *Pinus sylvestris* | 31 |
| Bhutan Pine | *Pinus wallichiana* | 31 |
| Western Yellow Pine | *Pinus ponderosa* | 31 |
| Atlas Cedar | *Cedrus atlantica* | 32 |
| Cedar of Lebanon | *Cedrus libani* | 32 |
| Deodar Cedar | *Cedrus deodara* | 33 |
| Common Juniper | *Juniperus communis* | 33 |
| European Larch | *Larix decidua* | 34 |
| Japanese Larch | *Larix kaempferi* | 34 |
| Hybrid Larch | *Larix x marschlinsii* | 34 |
| Maidenhair Tree | *Ginkgo biloba* | 35 |
| Italian Cypress | *Cupressus sempervirens* | 36 |
| Lawson Cypress | *Chamaecyparis lawsoniana* | 37 |
| Monterey Cypress | *Cupressus macrocarpa* | 37 |
| Leyland Cypress | *x Cupressocyparis leylandii* | 37 |
| Giant Redwood | *Sequoiadendron giganteum* | 38 |
| Monkey Puzzle | *Araucaria araucana* | 38 |
| Japanese Cedar | *Cryptomeria japonica* | 39 |
| Western Red Cedar | *Thuja plicata* | 39 |
| Incense Cedar | *Calocedrus decurrens* | 39 |
| Field Maple | *Acer campestre* | 42 |
| Japanese Maple | *Acer palmatum* | 42 |
| Norway Maple | *Acer platanoides* | 42 |
| Red Maple | *Acer rubrum* | 43 |

| | | |
|---|---|---|
| Sugar Maple | *Acer saccharum* | 43 |
| Sycamore | *Acer pseudoplatanus* | 43 |
| Sweetgum | *Liquidambar styraciflua* | 44 |
| White Poplar | *Populus alba* | 44 |
| London Plane | *Platanus x acerifolia* | 45 |
| Oriental Plane | *Platanus orientalis* | 45 |
| Buttonwood | *Platanus occidentalis* | 45 |
| Red Oak | *Quercus rubra* | 46 |
| Sessile Oak | *Quercus petraea* | 46 |
| Pedunculate Oak | *Quercus robur* | 47 |
| Scarlet Oak | *Quercus coccinea* | 47 |
| Turkey Oak | *Quercus cerris* | 47 |
| Hawthorn | *Crataegus monogyna* | 48 |
| Midland Hawthorn | *Crataegus laevigata* | 48 |
| Tulip Tree | *Liriodendron tulipifera* | 48 |
| Common Alder | *Alnus glutinosa* | 49 |
| Persian Ironwood | *Parrotia persica* | 49 |
| Downy Birch | *Betula pubescens* | 50 |
| Chinese Red-barked Birch | *Betula albosinensis* | 50 |
| Himalayan Birch | *Betula utilis* | 50 |
| Gray Alder | *Alnus incana* | 51 |
| Italian Alder | *Alnus cordata* | 51 |
| Snowy Mespil | *Amelanchier lamarckii* | 51 |
| Hackberry | *Celtis occidentalis* | 52 |
| Common Hornbeam | *Carpinus betulus* | 52 |
| Indian Bean Tree | *Catalpa bignonioides* | 52 |
| Paper-bark Birch | *Betula papyrifera* | 53 |
| Turkish Hazel | *Corylus colurna* | 53 |
| Pocket Handkerchief Tree | *Davidia involucrata* | 53 |
| Keyaki | *Zelkova serrata* | 54 |
| English Elm | *Ulmus procera* | 54 |
| Common Beech | *Fagus sylvatica* | 55 |
| Purple Beech | *Fagus sylvatica* Purpurea Group | 55 |
| Oriental Beech | *Fagus orientalis* | 55 |
| Black Mulberry | *Morus nigra* | 56 |
| White Mulberry | *Morus alba* | 56 |
| Mountain Snowdrop Tree | *Halesia monticola* | 56 |
| Common Crab Apple | *Malus sylvestris* | 57 |
| Siberian Crab Apple | *Malus baccata* | 57 |
| Cultivated Apple | *Malus domestica* | 57 |
| Antarctic Beech | *Nothofagus antarctica* | 58 |
| Roble Beech | *Nothofagus obliqua* | 58 |
| Chinese Necklace Poplar | *Populus lasiocarpa* | 59 |
| Apricot | *Prunus armeniaca* | 59 |
| Empress Tree | *Paulownia tomentosa* | 59 |
| Coigüe Southern Beech | *Nothofagus dombeyi* | 60 |
| Cork Oak | *Quercus suber* | 60 |
| Portuguese Laurel | *Prunus lusitanica* | 60 |

| | | |
|---|---|---|
| Soft Tree Fern | *Dicksonia antarctica* | 61 |
| Crack Willow | *Salix fragilis* | 62 |
| Golden Weeping Willow | *Salix x sepulcralis* 'Chrysocoma' | 62 |
| Common Osier | *Salix viminalis* | 62 |
| Tasmanian Blue Gum | *Eucalyptus globulus* | 63 |
| Cider Gum | *Eucalyptus gunnii* | 63 |
| Peach | *Prunus persica* | 64 |
| Tibetan Cherry | *Prunus serrula* | 64 |
| Willow-leaved Pear | *Pyrus salicifolia* | 64 |
| White Willow | *Salix alba* | 65 |
| Violet Willow | *Salix daphnoides* | 65 |
| Dragon's Claw Willow | *Salix babylonica var. pekinensis* 'Tortuosa' | 65 |
| Aspen | *Populus tremula* | 66 |
| Gray Poplar | *Populus x canescens* | 66 |
| Eastern Redbud | *Cercis canadensis* | 67 |
| Judas Tree | *Cercis siliquastrum* | 67 |
| Katsura Tree | *Cercidiphyllum japonicum* | 67 |
| American Basswood | *Tilia americana* | 68 |
| Small-leaved Linden | *Tilia cordata* | 68 |
| Large-leaved Linden | *Tilia platyphyllos* | 69 |
| Common Linden | *Tilia x europaea* | 69 |
| Killarney Strawberry Tree | *Arbutus unedo* | 72 |
| Madrone | *Arbutus menziesii* | 72 |
| Chilean Fire Bush | *Embothrium coccineum* | 72 |
| Common Holly | *Ilex aquifolium* | 73 |
| American Holly | *Ilex opaca* | 73 |
| Bay Laurel | *Laurus nobilis* | 73 |
| Olive | *Olea europaea* | 74 |
| Holm Oak | *Quercus ilex* | 74 |
| Chilean Myrtle | *Luma apiculata* | 75 |
| Bull Bay | *Magnolia grandiflora* | 75 |
| Pink Tulip Tree | *Magnolia campbellii* | 76 |
| Sprenger's Magnolia | *Magnolia sprengeri* | 76 |
| Saucer Magnolia | *Magnolia x soulangeana* | 76 |
| Star Magnolia | *Magnolia stellata* | 77 |
| Wilson's Magnolia | *Magnolia wilsonii* | 77 |
| Black Tupelo | *Nyssa sylvatica* | 77 |
| Wild Cherry | *Prunus avium* | 78 |
| Plum | *Prunus domestica* | 78 |
| Bird Cherry | *Prunus padus* | 78 |
| Sargent's Cherry | *Prunus sargentii* | 79 |
| Kanzan Flowering Cherry | *Prunus serrulata* 'Kanzan' | 79 |
| Great White Cherry | *Prunus* 'Taihaku' | 79 |
| Mount Fuji Cherry | *Prunus* 'Shirotae' | 80 |
| Blackthorn | *Prunus spinosa* | 80 |
| Common Pear | *Pyrus communis* | 80 |
| Whitebeam | *Sorbus aria* | 81 |
| Swedish Whitebeam | *Sorbus intermedia* | 81 |

# Glossary

A few terms can help you understand trees better and describe them with greater precision. Terms defined elsewhere in the glossary have been italicized.

**Acorn** Seed of the oak *genus* (*Quercus*) held in a rough-textured cup.

**Alternate** Describes leaves borne singly, in two vertical rows or spirally.

**Aril** A fleshy, often brightly colored, covering on a seed, such as on yews.

**Axil** The angle between two structures, such as the leaf and stem or the *midrib* and a small vein.

**Bract** A small, leaflike structure found at the base of flowers or in the *cone* of a conifer.

**Canopy** The top section of branches and leaves on a tree.

**Catkin** An unbranched and often drooping flower cluster of a single sex.

**Columnar** Taller than broad, with roughly parallel sides.

**Compound leaf** A leaf divided into *leaflets*.

**Cone** The flowering and fruiting structure of conifers.

**Conical** Widest at the bottom, tapering toward the top.

**Cultivar** A plant selection made by humans and maintained in cultivation.

**Deciduous** Describes a tree that is leafless for part of the year (usually winter).

**Dioecious** Describes trees that produce male and female flowers on separate trees.

**Down** Soft, hairlike covering often found on the surface of leaves.

**Elliptic** Describes a leaf that is widest at the middle and tapers relatively equally at both ends.

**Evergreen** Describes a tree that retains leaf cover throughout the year.

**Frond** Term given to the large divided leaves of ferns.

**Garden origin** Relates to plants that do not occur naturally in the wild and have been developed in cultivation.

**Genus (pl. Genera)** A category in classification consisting of a group of closely related species, and denoted by the first part of the scientific name, e.g. *Pinus* in *Pinus sylvestris*.

**Hermaphroditic** Trees, such as cherries, that produce flowers with both male and female reproductive organs.

**Hybrid** A cross between two different species.

**Lanceolate** Describes a lance-shaped leaf that widens just above the base before tapering toward the apex.

**Leading shoot** The shoot at the end of a main branch.

**Leaflet** One of the divisions that make up a *compound leaf*.

**Lenticel** A small pore on bark, shoots, and fruit through which air can pass.

**Lobe** A protruding part of a leaf, or sometimes a flower.

**Midrib** The primary, usually central, vein of a leaf or *leaflet*.

**Monoecious** Describes trees that bear separate male and female flowers on the same tree.

**Native** Occurring naturally in a particular region.

**Oblong** Describes a leaf that is longer than broad, and has parallel or nearly parallel sides.

**Obovate** Describes an egg-shaped leaf that is broadest above the middle.

**Opposite** Leaves borne in pairs on opposite sides of the stem.

**Ovate** Describes an egg-shaped leaf that is broadest toward the base.

**Ovule** Structure that contains the egg of a seed plant. After fertilization the ovule becomes the seed.

**Pealike** Describes a flower structure typical of members of the Pea (legume) family, with the *sepals* fused into a short tube, and usually with an erect upper petal, two wing petals, and two lower petals forming a keel.

**Petiole** The stalk of a leaf.

**Pollen** Spores or grains contained in a flower's anther that bear the male means of fertilization.

**Resin** An often sticky secretion of many plants, particularly conifer trees.

**Semidouble flower** A flower with more than the normal number of petals.

**Sepal** The usually green parts of a flower outside of the petals, collectively called the calyx.

**Simple leaf** A leaf that is not *compound* or divided.

**Sloe** The common name for the fruit of the blackthorn *Prunus spinosa*.

**Species** A classification category defining a group of similar and usually interbreeding plants, e.g. Scots Pine (*Pinus sylvestris*) is one species of the *Pinus* genus.

**Stalk** The part of a plant that attaches the leaf to the branch, also known as the *petiole*.

**Stamen** Male part of a flower, composed of an anther, normally borne on a stalk (filament).

**Style** Female part of a flower connecting the ovary and the stigma—the part that receives the pollen.

**Subspecies** A category of classification, below species, defining a group within a species, isolated geographically but able to interbreed with others of the same species.

**Sucker** Shoot arising from below the soil at the base of a tree.

**Tepal** *Sepals* and petals when they look alike.

**Terminal** Located at the end of a shoot, stem, or other organ.

**Variegated** Having more than one color; usually used to describe leaves, but sometimes flowers.

# Index

Page numbers in **bold** indicate main entry

# Acknowledgments

Dorling Kindersley would like to thank Taiyaba Khatoon and Ashwin Adimari for picture research.
The publisher would also like to thank the following for their kind permission to reproduce their photographs:
(**Key**: a-above; b-below/bottom; c-center; f-far; l-left; r-right; t-top)

**6-7 Corbis:** (cl). **12 Dorling Kindersley:** The Forestry Commission at Westonbirt, The National Arboretum (br). **13 Dorling Kindersley:** Batsford Arboretum (cr); The Royal Botanic Gardens, Kew (br). **16 Alamy Images:** shapencolour (cra). **20 Tony Russell:** (c). **23 Tony Russell:** (br, tr). **29 Tony Russell:** (br). **36 Dorling Kindersley:** (cr). **37 Dorling Kindersley:** Batsford Arboretum (br, bl). **39 Dorling Kindersley:** Batsford Arboretum (cr). **43 Dorling Kindersley:** Batsford Arboretum (cr). **44 Dorling Kindersley:** The Forestry Commission at Westonbirt, The National Arboretum (bc). **Tony Russell:** (t). **46 Dorling Kindersley:** Batsford Arboretum (br). **47 Dorling Kindersley:** Batsford Arboretum (br). **49 Dorling Kindersley:** Batsford Arboretum (br). **53 Dorling Kindersley:** The Forestry Commission at Westonbirt, The National Arboretum (cr). **54 Alamy Images:** Frank Blackburn (br). **Dorling Kindersley:** Batsford Arboretum (cr). **55 Dorling Kindersley:** The Forestry Commission at Westonbirt, The National Arboretum (br, cr). **57 Tony Russell:** (tr). **58 Dorling Kindersley:** The Royal Botanic Gardens, Kew (br). **59 Dorling Kindersley:** The Forestry Commission at Westonbirt, The National Arboretum (tr). **60 Dorling Kindersley:** Batsford Arboretum (cr); The Forestry Commission at Westonbirt, The National Arboretum (tr). **62 Tony Russell:** (br). **67 Dorling Kindersley:** (br); The Forestry Commission at Westonbirt, The National Arboretum (tr). **73 Dorling Kindersley:** The Forestry Commission at Westonbirt, The National Arboretum (cr). **76 Dorling Kindersley:** The Royal Botanic Gardens, Kew (cr). **77 Alamy Images:** Roger Phillips (cr). **Tony Russell:** (tc). **78 Dorling Kindersley:** Neil Fletcher (cr). **Tony Russell:** (tr). **84 Dorling Kindersley:** The Forestry Commission at Westonbirt, The National Arboretum (br); The Royal Botanic Gardens, Kew (cr). **85 Alamy Images:** (cr). **Dorling Kindersley:** Neil Fletcher (tr); The Forestry Commission at Westonbirt, The National Arboretum (cl). **86 Dorling Kindersley:** The Forestry Commission at Westonbirt, The National Arboretum (br); The Royal Botanic Gardens, Kew (tr). **87 Dorling Kindersley:** The Forestry Commission at Westonbirt, The National Arboretum (tr). **92 Dorling Kindersley:** Batsford Arboretum (tr). **96 Dorling Kindersley:** The Royal Botanic Gardens, Kew (br). **97 Corbis:** Martin B. Withers / Frank Lane Picture Agency (br). **98 Corbis:** Sally A. Morgan; Ecoscene (clb). **naturepl.com:** Simon Colmer (cla, cr). **99 Corbis:** VEM / Westend61 (tr). **naturepl.com:** Simon Colmer (cr). **Tony Russell:** (cla). **104 Dorling Kindersley:** Batsford Arboretum (c). **105 Dorling Kindersley:** The Forestry Commission at Westonbirt, The National Arboretum (bc). **Tony Russell:** (tr). **106 Dorling Kindersley:** (tr, br); Batsford Arboretum (crb, bl). **107 Dorling Kindersley:** The Forestry Commission at Westonbirt, The National Arboretum (crb, bc, br). **108 Dorling Kindersley:** Batsford Arboretum (cla, cl, clb, br); The Forestry Commission at Westonbirt, The National Arboretum (tc); The Royal Botanic Gardens, Kew (tr). **Tony Russell:** (bc). **109 Dorling Kindersley:** The Royal Botanic Gardens, Kew (clb). **110 Dorling Kindersley:** The Royal Botanic Gardens, Kew (bc). **111 Alamy Images:** Roger Phillips (tl). **Dorling Kindersley:** The Forestry Commission at Westonbirt, The National Arboretum (bl). **114 Tony Russell:** (crb). **115 Tony Russell:** (cra). **116 Alamy Images:** (cl). **Dorling Kindersley:** The Forestry Commission at Westonbirt, The National Arboretum (cla, crb, ca). **117 Dorling Kindersley:** Batsford Arboretum (cla). **118 Alamy Images:** Frank Blackburn (c). **Dorling Kindersley:** Batsford Arboretum (bc); The Royal Botanic Gardens, Kew (cl, bl); The Forestry Commission at Westonbirt, The National Arboretum (cra)

**Jacket images:** *Front:* **Alamy Images:** Arterra Picture Library (fbr), Clearvista Photography (fcr), garfotos (clb), Steffen Hauser / botanikfoto (fcra), Jaanus Järva (fcrb); *Back:* **Alamy Images:** Arco Images GmbH (cl), Nigel Cattlin (clb); **Corbis:** Martin B. Withers / Frank Lane Picture Agency (bl); *Spine:* **Alamy Images:** garfotos

All other images © Dorling Kindersley
For further information see: **www.dkimages.com**